Radar Imaging for Maritime Observation

Signal and Image Processing of Earth Observations Series

Series Editor

C.H. Chen

Radar Imaging for Maritime Observation

Edited by
Fabrizio Berizzi · Marco Martorella · Elisa Giusti

CRC Press
Taylor & Francis Group
Boca Raton London New York

CRC Press is an imprint of the
Taylor & Francis Group, an **informa** business

CRC Press
Taylor & Francis Group
6000 Broken Sound Parkway NW, Suite 300
Boca Raton, FL 33487-2742

First issued in paperback 2017

ISBN 13: 978-1-138-49089-5 (pbk)
ISBN 13: 978-1-4665-8081-7 (hbk)

Library of Congress Cataloging-in-Publication Data

Names: Berizzi, Fabrizio, editor. | Martorella, Marco, editor. | Giusti, Elisa, editor.
Title: Radar imaging for maritime observation / editors, Fabrizio Berizzi, Marco Martorella, and Elisa Giusti.
Description: Boca Raton, FL : Taylor & Francis, 2016. | Series: Signal and image processing of earth observations | Includes bibliographical references and index.
Identifiers: LCCN 2016003199 | ISBN 9781466580817 (alk. paper)
Subjects: LCSH: Oceanography--Remote sensing. | Synthetic aperture radar. | Inverse synthetic aperture radar.
Classification: LCC GC10.4.R4 R37 2016 | DDC 551.46028/4--dc23
LC record available at https://lccn.loc.gov/2016003199

Visit the Taylor & Francis Web site at
http://www.taylorandfrancis.com

and the CRC Press Web site at
http://www.crcpress.com

Dedication

To my family for their daily, moral support.
Fabrizio Berizzi

To my wife Petrina and my children Aisha, Tiago, and Estelle
for their love, which gives me the strength and the courage to face
the challenges of life.
Marco Martorella

To my loving parents, my husband Andrea, and my daughter
Giada for their unfailing love, loyalty, and support.
Elisa Giusti

Contents

PART I   SAR and ISAR Signal Processing

M. Martorella, D. Cataldo, S. Gelli, and S. Tomei

PART II   Applications

M. Martorella, F. Berizzi, E. Giusti, and A. Bacci

M. Martorella, F. Berizzi, E. Giusti, C. Moscardini, A. Capria, D. Petri, and M. Conti

List of Figures

List of Tables

Preface

The term maritime observation has been deliberately chosen to encapsulate both the concept of remote sensing of the sea surface and maritime surveillance. The two scenarios are significantly different in terms of applications related to them although they share a fundamental common ingredient, which is the sea surface. The presence of the sea surface characterizes uniquely the radar echoes and the way they must be processed either when some sea surface parameters must be estimated or when targets must be detected or identified. The use of radar for maritime observation finds its roots in the early days of radar when radars were installed on ships as support for the navigation and for maritime surveillance purposes. In fact, the ability of radar to operate in all weather/all day conditions made it appealing to a number of applications, from civil to military scenarios.

The interaction of radar microwave signals with the sea surface is very complex and it requires a good understanding of basic physics of backscattering phenomena to maximize the information that can be extracted. Theoretical and empirical studies have been conducted for decades with the aim to understand such physical phenomena and consequently improve detection, estimation and classification algorithms. With the advent of high spatial resolution radar and subsequently synthetic aperture radar (SAR), radar images became available that could be used also in the area of maritime observation. Nowadays, two-dimensional (2D) radar imaging is recognized as one of the most important tools for monitoring the sea surface. As a matter of fact, a large number of airborne and spaceborne platforms for Earth observation are equipped with radar imaging systems. Moreover, some coastal and ship-borne radars have radar imaging capability to improve the maritime situational awareness.

Motivated by these facts and based on the experience of the Radar Laboratory of the University of Pisa and of the Radar and Surveillance System (RaSS) national laboratory of the National Interuniversity Consortium of Telecommunication (CNIT), we decided to propose this book with the aim to present the most recent results in radar imaging for maritime observation. This book is intended both for beginners and for experts in this field as it treats the theoretical aspects as well as the applications.

The book is organized into two sections. The first section (SAR and ISAR image processing) contains the conceptual and theoretical aspects of both SAR and ISAR. More specifically, standard SAR/ISAR and novel imaging techniques, including bistatic ISAR, passive ISAR (P-ISAR) and the 3D Interferometric ISAR (3D InISAR) are detailed. The second section (Applications) focuses on the use of such techniques for maritime observation. Results based on the use of real data and related to scenarios of interest are presented throughout this section.

Section I is composed of six chapters. Chapter 1 introduces the principle of radar imaging and specifically the concepts of high range and cross-range resolutions, the relationship between SAR and ISAR and a generic received signal model that will be used throughout the remainder of the book. Chapter 2 is focused on SAR. The SAR signal model is first defined and the main SAR image formation techniques known in the literature are described. Chapter 3 refers to ISAR. The main steps of the ISAR image processing chain, namely image formation, time windowing, motion compensation and image scaling are detailed. A mention of ISAR parameter setting for ISAR system design is also made. Chapter 4 introduces the new concept of bistatic ISAR (B-ISAR). The bistatic geometry and modeling is introduced first, then bistatic image formation is introduced by making use of the equivalent monostatic configuration. The distortions caused by the bistatic geometry are also analyzed directly using a mathematically derived ISAR point spread function. The behavior of B-ISAR in the presence of synchronization errors is also dealt with. Chapter 5 concerns the use of passive radar for target imaging. In particular, the signal pre-processing chain for the formation of range Doppler maps is described by making use of the batch algorithm. Passive ISAR (P-ISAR) theory and signal modeling is then introduced. Reconstruction of P-ISAR images is illustrated together with a performance analysis. Chapter 6 regards 3D interferometric ISAR (3D-InISAR), where the height of each single target scatterer point is reconstructed in order to have a 3D plot of objects. A multi-channel ISAR signal model is defined to derive the main InISAR signal processing. The use of multiple interferometric radars also allows for the estimation of the image plane orientation and the effective rotation vector for cross-range scaling. Analysis of the performance is provided through the use of suitable parameters.

Section II is organized in five chapters, each containing applications of maritime observation by means of radar images. Chapter 7 is devoted to the detection of ships from SAR images. Image cleaning, several detectors and post-processing techniques are included. Three different case studies are shown to demonstrate the validity of the proposed techniques. Chapter 8 refers to the detection of oil spills with SAR images. Fractal modeling and analysis of SAR images is used for oil spill detection and discrimination to other look-alike dark areas. Results based on the use of real spaceborne SAR images are presented to show the effectiveness of the approach. Chapter 9 concerns the application of ISAR processing to refocus moving targets in SAR images. Real cases related to maritime scenarios are shown to demonstrate the capability of the proposed techniques. Chapter 10 is an interesting application of passive ISAR for harbor surveillance and protection. Two case studies are proposed to see how this processing can be applied to real scenarios. Chapter 11 deals with the application of 3D InISAR from a radar network to reconstruct the 3D shape and to extract features such as the size of a ship for traffic monitoring and safe navigation in internal port waters or along the coastline.

Acknowledgments

We express our sincere thanks to Professor Enzo Dalle Mese for having doggedly pursued his goal over the years to create a national radar research laboratory in Pisa. It is thanks to his work and guidance that the Italian National Laboratory of Radar and Surveillance Systems (RaSS) has been created and has become a leading research institution in the radar and surveillance systems field. His perseverance, motivation, and knowledge have laid the foundation of our research group.

We also thank our colleagues who were not directly involved in writing this book although they have contributed, with their work, to the prestige that our laboratory has gained in these recent years.

We are also grateful to the Italian Space Agency (ASI), the European Defence Agency (EDA), Italian Ministry of Defence (MoD) and the Italian Ministry of Education University and Research (MIUR) for having partially funded our research studies and the ASI for having provided real COSMO SkyMed images in the framework of the first announcement of opportunity for the exploitation of COSMO SkyMed products, which have been used in this book to show examples of SAR/ISAR processing. We also acknowledge the CSSN-ITE "Istituto Vallauri"of the Italian Navy and the NATO CSO for supporting both logistically and economically some of the measurement campaigns that have been run for collecting real data, which have been used to produce results shown in this book.

Contributors

Alessio Bacci
Department of Information
   Engineering, University of Pisa
Radar and Surveillance System
   (RaSS) National Laboratory,
   National Inter-University
   Consortium for
   Telecommunications (CNIT)
Pisa, Italy

Fabrizio Berizzi
Department of Information
   Engineering, University of Pisa
Radar and Surveillance System
   (RaSS) National Laboratory,
   National Inter-University
   Consortium for
   Telecommunications (CNIT)
Pisa, Italy

Amerigo Capria
Radar and Surveillance System
   (RaSS) National Laboratory,
   National Inter-University
   Consortium for
   Telecommunications (CNIT)
Pisa, Italy

Davide Cataldo
Department of Information
   Engineering, University of Pisa
Radar and Surveillance System
   (RaSS) National Laboratory,
   National Inter-University
   Consortium for
   Telecommunications (CNIT)
Pisa, Italy

Michele Conti
Radar and Surveillance System
   (RaSS) National Laboratory,
National Inter-University
   Consortium for
   Telecommunications (CNIT)
Pisa, Italy

Samuele Gelli
Department of Information
   Engineering, University of Pisa
Radar and Surveillance System
   (RaSS) National Laboratory,
   National Inter-University
   Consortium for
   Telecommunications (CNIT)
Pisa, Italy

Elisa Giusti
Radar and Surveillance System
   (RaSS) National Laboratory,
   National Inter-University
   Consortium for
   Telecommunications (CNIT)
Pisa, Italy

Stefano Lischi
Department of Information
   Engineering, University of Pisa
Radar and Surveillance System
   (RaSS) National Laboratory,
   National Inter-University
   Consortium for
   Telecommunications (CNIT)
Pisa, Italy

Alberto Lupidi
Radar and Surveillance System
   (RaSS) National Laboratory,
   National Inter-University
   Consortium for
   Telecommunications (CNIT)
Pisa, Italy

Marco Martorella
Department of Information
Engineering, University of Pisa
Radar and Surveillance System
(RaSS) National Laboratory,
National Inter-University
Consortium for
Telecommunications (CNIT)
Pisa, Italy

Christian Moscardini
Radar and Surveillance System
(RaSS) National Laboratory,
National Inter-University
Consortium for
Telecommunications (CNIT)
Pisa, Italy

Dario Petri
Radar and Surveillance System
(RaSS) National Laboratory,
National Inter-University
Consortium for
Telecommunications (CNIT)
Pisa, Italy

Federica Salvetti
Department of Information
Engineering, University of Pisa
Radar and Surveillance System
(RaSS) National Laboratory,
National Inter-University
Consortium for
Telecommunications (CNIT)
Pisa, Italy

Daniele Staglianò
Department of Information
Engineering, University of Pisa
Radar and Surveillance System
(RaSS) National Laboratory,
National Inter-University
Consortium for
Telecommunications (CNIT)
Pisa, Italy

Sonia Tomei
Department of Information
Engineering, University of Pisa
Radar and Surveillance System
(RaSS) National Laboratory,
National Inter-University
Consortium for
Telecommunications (CNIT)
Pisa, Italy

List of Symbols

$A_V[\cdot]$	Average operator
A_R	Surveillance channel in passive radar
A_{ref}	Reference channel in passive radar
B	Signal bandwidth
B_L	Radar baseline
B_S	Italian DVB-T signal instantaneous bandwidth
c	Light speed in a vacuum
$C_{s_T}(t)$	Autocorrelation function of the transmitted signal
cr_s	Doppler shift in the bistatic configuration
d_H	Horizontal baseline in 3D ISAR
d_V	Vertical baseline in 3D ISAR
D_{x_1}	Cross-range non-ambiguity region
D_{x_2}	Range non-ambiguity region
$D(\nu)$	Chirp-like distortion term in the bistatic configuration
$E[\cdot]$	Statistical expectation operator
f	Range frequency
f_0	Carrier frequency
$f(\mathbf{x})$	Target reflectivity function
$f_B(\mathbf{x})$	Bistatic target reflectivity function
$f(x_1, x_2)$	Target reflectivity function projection onto the (x_1, x_2) plane
$f_D(t)$	Doppler frequency as a function of the continuous slow time
h	Platform height
$h(t)$	Matched filter impulse response
$\mathbf{i}_{LoS}$	LoS unit vector
$\mathbf{i}_{LoS,T}$	LoS unit vector TX
$\mathbf{i}_{LoS,R}$	LoS unit vector RX
$\mathbf{i}_{LoS,B}$	LoS unit vector (bistatic)
$\mathbf{i}_{CR}$	Cross-range unit vector (bistatic)
$\tilde{\mathbf{i}}_{CR}$	Orthogonalized cross-range unit vector (bistatic)
$\mathbf{i}_{x_1}$	Unit vector identifying the x_1 axis direction
$\mathbf{i}_{x_2}$	Unit vector identifying the x_2 axis direction
$\mathbf{i}_{x_3}$	Unit vector identifying the x_3 axis direction
$\mathbf{i}_\Omega$	Unit vector identifying the unit effective rotation vector
$I(\cdot, \cdot)$	ISAR image
$IF(\tau, \nu)$	Improvement factor in passive radar
K	Number of scatterers composing the target
$K(n)$	Phase modulation term (bistatic geometry)
K_0	Phase modulation term $t = 0$ (bistatic geometry)
$ln(\cdot)$	Natural logarithm

L_{cr}, L	Antenna size along the cross-range direction
M_f	Number of transmitted frequencies
M	Number of delay-time samples
n	Pulse index
n_B	Number of batches
N	Pulse samples
N_ν	Number of Doppler bins
N_w	Time window length in number of sweeps
n_w	Time window position
O	Focusing point on the target
$p(p, \theta, 0)$	Cylindrical coordinates of the target
$q(t)$	Continuous *rect* function in passive radar
$\mathcal{P}_{ref}(f)$	Energy spectral density
$\hat{\mathcal{P}}_{ref}(f)$	Estimated energy spectral density
$R_0(t)$	Radar-target distance
$\mathbf{R}_t$	Global rotation matrix in 3D InISAR
$\mathbf{R}_\mu$	Yaw rotation matrix
$\mathbf{R}_\nu$	Pitch rotation matrix
$\mathbf{R}_\phi$	Roll rotation matrix
R_{Rx}	Radial distance between the target and the Rx
R_{Tx}	Radial distance between the target and the Tx
$s_T(t)$	Transmitted signal
$s_R(t)$	Received signal
$s(t)$	Matched filter output signal
$SNR_b(\tau, \nu)$	Signal-to-noise ratio in a single batch for passive radar
$SNR_{opt}(\tau, \nu)$	Signal-to-noise ratio in the optimum case for passive radar
$SNR(\tau, \nu)$	Signal-to-noise ratio in passive radars
$s_{ref}(t)$	Reference signal in passive radars
$s_{refb}(t)$	Reference signal in a batch
$s_{Rb}(t)$	Received or surveillance signal in a batch
$S_a(\tau, n)$	Roughly aligned range profiles
$S(f, n)$	SAR signal at the output of the matched filter in the frequency/slow time domain
$S_0(n)$	$S_a(\tau, n)\|_{\tau=\tau_0}$
$S_m(n)$	$S_a(\tau, n)\|_{\tau=\tau_m}$
$S_0, c(n)$	Phase compensated range/slow-time signal relative to the range cell τ_0
$S_{2D-INT}(X_1, X_2)$	Signal at the output of the 2D interpolation step
$S_c(\tau, n)$	Phase compensated range/slow-time signal
$S_T(f)$	Transmitted signal in the frequency/slow-time domain
$S_C(f, n)$	Motion compensated received signal
$S_{CR}(\tau, n)$	Received signal after range compression and motion compensation in the delay time/slow time domain
$S_{ref}(\cdot, \cdot)$	Reference signal

T_B	Batch length
T_I	Pulse duration
T_{int}	Integration time in passive radars
T_ξ	Reference system embedded on the radar
T_y	Reference system embedded on the target
T_x	Reference system embedded on the target (fixed)
T_w	Temporal window length
f	Fast time
T_R	Pulse repetition interval
T_{obs}	Observation time
T_w	Window length
u	Platform position on the trajectory
V	Volume where the target reflectivity is defined
v	Platform velocity in SAR systems or target velocity in ISAR systems
v_{rp}	Radial velocity of target (modulus)
$w(\cdot,\cdot)$	Point spread function in the delay time/Doppler domain
$W(f,n)$	Frequency/slow time domain in which the signal is defined
$\tilde{w}(\tau,n)$	Point spread function in the delay time/Doppler domain
$W_F(f)$	Signal support in the frequency domain
$W_T(n)$	Signal support in the slow time domain
(X_1, X_2)	Spatial frequencies
x_1	Cross-range direction
x_2	Range direction
$\mathbf{x}_k$	Location of the kth scatterer
$\boldsymbol{\alpha}$	Target motion parameters
$\beta(t)$	Bistatic angle
$\delta(\cdot)$	Dirac Delta function
Δf	Frequency spacing
$\Delta\vartheta$	Aspect angle variation in a CPI
δ_τ	Delay time resolution
δ_ν	Doppler frequency resolution
$\delta_{cr} = \delta_{x_1}$	Cross-range resolution
$\delta_r = \delta_{x_2}$	Range resolution
δ_T	Pulse duration at the output of the matched filter
$\Delta X_1/\Delta X_1$	Spacial frequencies spacing
$\Delta\phi$	Phase gradient
$\Delta\hat{\phi}$	Phase gradient estimate
Δ_{x_1}	Cross-range non-ambiguity region
Δ_{x_2}	Range non-ambiguity region
ϑ	Aspect angle
ϑ_{cr}	Antenna angular resolution
ϑ_{in}	Incidence angle
ϑ_{az}	Azimuth aperture

λ	Wavelength
μ	Chirp rate
ν	Doppler frequency
ρ	Compression factor
τ	Round-trip delay
σ_k	Reflectivity of the kth target scatterer
$\phi(f, n)$	Phase history in the frequency/slow-time domain
$\phi(n)$	Slow-time phase history
$\hat{\phi}(n)$	Estimate of the slow-time phase history
$\chi(\tau, \nu)$	Range Doppler (cross) ambiguity function
$\chi_n(\tau, \nu)$	Range Doppler (cross) ambiguity function in the nth batch
$\chi_{opt}(\tau, \nu)$	Optimum range Doppler (cross) ambiguity function
$\chi_{cc}(\tau, n)$	Signal in the range/slow-time domain after cross-correlation
$\chi^{(s)}(\tau, \nu)$	Target contribution at the output of the CAF algorithm in passive radars
$\chi^{(w)}(\tau, \nu)$	Noise contribution at the output of the CAF algorithm in passive radars
$\chi_b^{(s)}(\tau, \nu)$	Target contribution at the output of the CAF algorithm in passive radars
$\chi_{opt}^{(s)}(\tau, \nu)$	Target contribution at the output of the optimum algorithm in passive radars
$\boldsymbol{\Omega}_T(t)$	Total rotation vector
Ω	Effective rotation vector
Ω_B	Bistatic effective rotation vector
Ω_{BT}	Total bistatic effective rotation vector
$\omega_a(t)$	LoS azimuth angle with respect to the reference system embedded on the target
$\omega_e(t)$	LoS elevation angle with respect to the reference system embedded on the target

List of Abbreviations

2D	Two-dimensional
2D-INT	Two-dimensional interpolation
3D	Three-dimensional
AF	Ambiguity function
ARMA	Auto regressive moving average
ATC	Automatic target classification
ATR	Automatic target recognition
B-ISAR	Bistatic inverse synthetic aperture radar
B-IPP	Bistatic image projection plane
BEM	Bistatically equivalent monostatic
BP	Back projection
CAF	Cross ambiguity function
CA-CFAR	Cell average constant false alarm rate
CFAR	Constant false alarm rate
COFDM	Coded orthogonal frequency division multiplexing
CPI	Coherent processing interval
CS	Chirp scaling
DAB	Digital audio broadcasting
DFT	Discrete Fourier transform
DRD	Differential range deramp
DSA	Dominant scatterer autofocus
DVB-T	Digital video broadcasting-terrestrial
DVB-S	Digital video broadcasting-satellite
DVB-SH	Digital video broadcasting-satellite to handheld
ECM	Electronic counter measure
EIRP	Equivalent isotropic radiation power
e.m.	Electromagnetic
ESD	Energy spectral density
FARIMA	Fractionally integrated autoregressive moving average
FEXP	Fractionally EXPonential
FJ	Frequency jitter
FFT	Fast Fourier transform
FM	Frequency modulation
FMCW	Frequency modulated continuous waveform
FT	Fourier transform
GEV	Generalized extreme value
GSM	Global system for mobile communications
HS	Hot spot
IC	Image contrast

ICBA	Image contrast based algorithm
IE	Image entropy
IEBA	Image entropy based algorithm
IF	Improvement factor
IFT	Inverse Fourier transform
IFFT	Inverse fast Fourier transform
IPP	Image projection plane
IO	Illuminator of opportunity
ISAR	Inverse synthetic aperture radar
InISAR	Interferometric ISAR
LFM	Linear frequency modulated
LoS	Line of sight
LPFT	Local polynomial Fourier transform
LRD	Long range dependence
LSE	Least square error
MC-CLEAN	Multi-channel CLEAN
M-ICBA	Multi-channel image contrast based algorithm
MF	Matched filter
ML	Maximum likelihood
MMSE	Minimum mean square error
MV	Minimum variance
MRPSD	Mean radial power spectral density
MSA	Multiple scatterer algorithm
MVA	Minimum variance algorithm
NCTR	Non-cooperative target recognition
OFDM	Orthogonal frequency division multiplexing
PB-ISAR	Passive bistatic ISAR
PF	Polar format
PFT	Polynomial Fourier transform
PGA	Phase gradient autofocus
PLS	Probabilistic least square
PR	Passive radar
PRF	Pulse repetition frequency
PRT	Pulse repetition interval
PSF	Point spread function
PSD	Power spectral density
PPP	Prominent point processing
QAM	Quadrature amplitude modulation
RCMC	Range cell migration correction
RD	Range-Doppler
RID	Range instantaneous Doppler
RMSE	Root mean square error
SAR	Synthetic aperture radar
SNR	Signal-to-noise ratio

SRC	Secondary range compression
SRD	Short range dependence
TFT	Time Fourier transform
TFD	Time frequency distribution
TPS	Transport parameter signaling
UMTS	Universal mobile communication systems
W-CFAR	Wavelet constant false alarm rate
WiMAX	WiFi and worldwide interoperability for microwave access

Part I

SAR and ISAR Signal Processing

1 Principles of Radar Imaging

F. Berizzi, M. Martorella, and E. Giusti

CONTENTS

Synthetic aperture radar (SAR) and inverse synthetic aperture radar (ISAR) are referred to in literature as imaging radars because of their ability to reconstruct electro-magnetic (e.m.) images of natural and man-made objects by coherently processing the echoes coming from the targets at different aspect angles. The comparison between radar and photographic images is quite common in literature, since both systems perform a transformation that maps a 3D object to a 2D space. However, differences exist which concern the mapping transformation and the image feature. While the latter is quite obvious since different imaging systems use different illuminators thus producing images representing different characteristics of the target, the former is of more difficult interpretation. Differently from electro-optic systems, where the image projection plane (IPP), which is the 2D plane which a 3D target is projected onto, coincides with the focal plane of the sensor, for an imaging radar, the IPP depends on the relative motion between the radar and the target. Therefore, the imaging radar IPP can be a 2D plane arbitrarily oriented in the 3D space, which depends on the complexity of the target motions with respect to the radar.

As well as other types of images, radar images are usually characterized by some quality indexes, such as geometrical resolutions, radiometric resolution, signal-to-noise ratio, etc. Among them, the geometrical resolution plays a key role since the finer the spatial resolutions of the system, the more the target

3

details that can be observed in the image. As a consequence, high resolution radar images allow for a proper understanding of the observed target thus opening the doors to automatic target classification (ATC) and automatic target recognition (ATR).

Besides this brief introduction to radar imaging systems, the purpose of this section is to introduce the fundamental concepts of radar imaging theory, being the radar image spatial resolutions, as well as the key aspects that differentiate SAR systems from ISAR systems. Specifically, these are the objectives of Section 1.1 and 1.2, respectively. Finally, a signal model and the notation that will be used throughout the book will be introduced in Section 1.3.

It is worth mentioning that, for a better comprehension of the book content, the reader should be familiar with the basic concept of pulse radars.

1.1 HIGH RESOLUTION RADARS

Spatial resolutions represent a key feature for a good understanding of the imaged target, thus allowing ATC and ATR techniques to be successfully applied. Before going through the spatial resolution definition, we briefly recall the concept of radar image. A radar image can be defined as the spatial distribution of the e.m. reflectivity of an object mapped onto a two-dimensional plane. The target reflectivity is a measure of the electric intensity field backscattered by the target and collected by the radar. The backscattering effect is caused by the e.m. field irradiated by the radar that induces electric surface currents on the illuminated target, which in turn produce an e.m. field that partially propagates back to the radar. Therefore, the imaging radar measures the spatial distribution of the target reflectivity function projected onto a two-dimensional plane. As mentioned before, the finer the spatial resolutions, the more detailed the target radar image which means that different parts of the target corresponding to different reflectivity features can be identified.

From a radar perspective, the IPP is identified by the radial or range direction, which coincides with the antenna bore-sight, and by the cross-range direction that lies in the plane orthogonal to the range direction and whose orientation depends on the radar-target geometry and target motion. Therefore, fine spatial resolutions are needed in both range and cross-range directions to enable radar imaging capabilities in a radar system.

The resolution can be generally defined as the minimum distance between two alike quantities such that a system is able to distinguish them as separated from its measurements. This general definition may be applied to all sorts of measurements and systems and can be measured in terms of meters for spatial resolutions, Hz for frequency resolution, seconds for time resolution, and so on.

By applying this definition to a radar system, the resolution is defined as *the minimum distance along a pre-defined direction between two point-like scatterers with equal reflectivity such that they appear as separated objects in*

Figure 1.1 SAR/ISAR system

the radar image or, in other words, such that they can be distinguished by the radar system.

The resolution depends on the system and, specifically, it is determined by the system point spread function (PSF), which is interpreted as the impulse response of the radar system, coinciding with reconstructed radar imaging function of a unitary point scatterer. For SAR and ISAR systems, for example, the impulse response is usually approximated by means of a two-dimensional *sinc*-like function. By assuming the target composed of independent point-like scatterers, the SAR/ISAR image can be interpreted as the convolution between the 3D target's reflectivity function projected onto the IPP and the SAR/ISAR impulse response, as illustrated in Figure 1.1. Under the hypothesis of independent point-like scatterers, the target reflectivity function can be assumed to be a superposition of δ-like functions, then the SAR/ISAR image will be the superposition of *sinc*-like functions centered at the scatterers coordinates in the IPP plane.

It is clear that the sharper the main lobe of the *sinc*-function, the better the capability of the system to separate the scatterers contributions.

By dealing with two-dimensional radar images, two types of resolutions are defined, namely the range resolution and the cross-range resolution. As obvious, the range resolution is the capability of the system to distinguish two scatterers along the range direction, while the cross-range resolution is the capability of the system to distinguish two scatterers along the cross-range direction.

1.1.1 HIGH RANGE RESOLUTION

In narrow band pulse radars, the range resolution is usually linked to the pulse duration, T_I . In fact, an echo from a point-like scatterer persists for T_I seconds. As a consequence, the echo relative delay for a close range scatterer must be at least T_I to be distinguished.

This leads to the definition of the range resolution as follows:

$$\delta_r = \frac{cT_I}{2} \qquad (1.1)$$

where c is the light speed in a vacuum and approximately corresponds to the e.m. wave propagation speed. From Equation (1.1), it is clear that to improve the range resolution, the pulse duration should be reduced. This however implies a higher peak power to obtain unchanged performance in terms of probability of detection and false alarm.

To avoid the trade-off between range resolution and transmitted power peak, pulse compression theory was formulated [7, 10, 4, 1]. The pulse compression permits transmitting a signal with pulse duration sufficiently long to limit the transmitted peak power and, at the same time, to get finer range resolutions.

The pulse compression is realized by means of the *matched filter* (MF).

The MF ensures the maximum *signal-to-noise ratio* (SNR), at its output around the target delay. In fact, the *compression* ensures that the energy of the transmitted pulse is more concentrated around the target delay, thus leading to a higher SNR.

The MF calculates the cross-correlation function of the received signal with the reverse conjugate of the transmitted signal as follows:

$$s(t) = s_R(t) \otimes s_T^*(-t) \tag{1.2}$$

where $s_R(t)$ is the signal at the input of the filter and $h(t) = s_T^*(-t)$ is the impulse response of the MF. As for radar applications, the received signal is supposed to be a delayed replica of the transmitted signal, $s_T(t)$, then $s(t)$ is also proportional to the autocorrelation function $C_{s_T}(t)$ of the transmitted signal, apart from a time delay.

Therefore, the Fourier transform (FT) of the output is proportional to the energy spectral density (ESD) of the transmitted signal, that is:

$$ESD(f) \propto FT\left[s_T(t) \otimes s_T^*(-t)\right] = |S_T(f)|^2 \tag{1.3}$$

where $S_T(f) = FT\left[s_T(t)\right]$ with $FT\left[\cdot\right]$ the Fourier transform operator. By exploiting the uncertainty relationship of a Fourier transform pair, it follows that the wider the bandwidth of the transmitted bandwidth, the shorter the time duration of $s(t)$.

In other words, the so-called time-bandwidth product of the output signal is constant. This property is summarized in Equation (1.4)

$$B \cdot \delta_\tau = \mathcal{H} \tag{1.4}$$

where the value of $\mathcal{H}$ depends on the waveform of the transmitted signal, B is the bandwidth of the transmitted signal and δ_τ is the pulse duration at the output of the MF. For a linear frequency modulated (LFM) signal, $\mathcal{H} \simeq 1$, which is a quite high value with respect to other waveforms, and this makes such modulated pulses very attractive for radar applications.

It is clear that two separate scatterers can be distinguished if their relative echoes are separated at least by the pulse duration at the output of the MF, that is δ_τ. By exploiting Equation (1.1), the range resolution can be found equal to:

$$\delta_r = \frac{c\delta_\tau}{2} = \frac{c\mathcal{H}}{2B} \tag{1.5}$$

The joint use of wide-band transmitted signals and matched filter techniques realizes the *pulse compression*. The term *pulse compression* derives from the fact that the time duration of a pulse is shortened at the output of the MF by a factor $\rho = \frac{T_I}{\delta_\tau}$, namely the *compression factor*. It is worth remarking that wide bandwidth transmitted signal is a necessary condition to have a compression gain, i.e., $\rho > 1$. Phase and/or frequency modulation must be used to have effective compression.

Therefore, the *pulse compression* allows high range resolution to be achieved by transmitting suitable wide bandwidth signals with unchanged radar range coverage area.

1.1.2 HIGH CROSS-RANGE RESOLUTION

The ability to resolve two scatterers in the cross-range direction is related to their angular separation in the plane (x_1, x_2), which identifies the IPP. By referring to the monostatic radar geometry in Figure 1.2, the cross-range direction x_1 is orthogonal to the radial direction x_2. Traditionally angular resolution was achieved by using large antennas, as the antenna beam-width is inversely related to the antenna size along x_1. For example, the angular resolution of a rectangular antenna can be approximately calculated as follows:

$$\vartheta_{cr} \simeq \frac{\lambda}{L_{cr}} \tag{1.6}$$

where $\lambda = \frac{c}{f_0}$ is the wavelength, L_{cr} is the antenna size along the cross-range direction and ϑ_{cr} is the angular resolution in the the cross-range/range plane.

The same concept applies for the elevation direction.

The antenna beam-width, and therefore the angular resolution, is however not sufficient to provide fine cross-range resolution, either because it refers to the angular domain and not to the spatial domain (images should be scaled from angular to spatial coordinates) or because the cross-range resolution is also a function of the target range as shown in Equation (1.7)

$$\delta_{x_1} \simeq R_0 \vartheta_{cr} = \frac{R_0 \lambda}{L_{cr}} \tag{1.7}$$

It is also evident that to get a fine cross-range resolution at long ranges, wider antennas should be used. However, there is a practical limitation for the antenna size.

Another solution is to use an antenna array, which may relax this problem. However, even in the case of antenna array, a large number of antennas should be used to form a long array.

SAR and ISAR technologies have been proposed in literature to overcome such limitations [4, 8, 5, 1, 3, 6]. Both these techniques are based on the

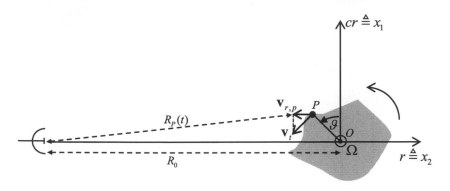

Figure 1.2 Geometry definition

synthesis of a long array by exploiting the relative motion between the radar platform and the target. The relative motion causes a variation of the aspect angle over time. The reason why an aspect angle variation over time leads to a fine enough cross-range resolution is explained below.

Let us refer again to the geometry in Figure 1.2, where the radar illuminates a target that rotates with a rotational velocity vector $\mathbf{\Omega}$ orthogonal to the (x_1, x_2) plane. Let us consider now a point on the target with cylindrical coordinates $\mathbf{p} = (p, \vartheta, 0)$. Because of the target motion this point has a radial velocity with respect to the radar equal to $v_{r_P} = v_t \cdot cos(\vartheta)$, where v_t is the modulus of the tangential velocity of the point. The tangential velocity is linked to the rotation vector by the following formula:

$$\mathbf{v}_t = \mathbf{\Omega} \times \mathbf{p} \tag{1.8}$$

where $\times$ represents the cross-product operator. Equation (1.8) can be rewritten as follows:

$$v_{r_P} = \Omega \cdot p \cdot cos(\vartheta) = \Omega \cdot x_{1_P} \tag{1.9}$$

where $\Omega = |\mathbf{\Omega}|$ and $x_{1_P} = p \cdot cos(\vartheta)$ represents the cross-range coordinate of the point in the Cartesian reference system $T_X(x_1, x_2, x_3)$.

Assuming that the radar transmits a pure tone at frequency f_0, i.e. $s_T(t) = cos(2\pi f_0 t)$, the received signal can be written taking into account the round-trip time delay, $\tau_P(t)$, that is

$$s_R(t) = cos(2\pi f_0(t - \tau_P(t))) \tag{1.10}$$

where $\tau_P(t) = \frac{2R_P(t)}{c} \simeq \frac{2}{c}(R_0 + v_{r_P}t)$. By substituting $\tau_P(t)$ into Equation (1.10), we obtain:

$$s_R(t) = cos(2\pi (f_0 + \nu_P) t + \phi_0) \cdot rect\left(\frac{t}{T_{obs}}\right) \tag{1.11}$$

where $\phi_0 = -\frac{4\pi R_0}{\lambda}$, $\nu_P = -\frac{2v_{rP}}{\lambda}$ is the Doppler frequency of the point P, $rect(x) = 1$ when $|x| < 0.5$ and 0 otherwise, and T_{obs} refers to the observation time.

Since the radial velocity $v_{r,P}$ is related to the point cross-range coordinate x_{1P}, it is obvious that the received signal contains information about the cross-range resolution.

By inverse Fourier transforming the signal in Equation (1.11), the signal in the Fourier domain is:

$$S_R(\nu) = T_{obs} \cdot sinc\left(T_{obs}\left(\nu - \nu_P\right)\right) \tag{1.12}$$

where ν is the Doppler frequency variable. Equation (1.12) shows that a *sinc*-like function is obtained in the Doppler frequency coordinate centered at the Doppler frequency of the scatterer (ν_P). The Doppler resolution is given by $\delta_\nu = \frac{1}{T_{obs}}$ and corresponds approximatively to the -3 dB width of the *sinc* main lobe.

By exploiting the relationship between the radial velocity, v_{r_P}, and the cross-range coordinate, Equation (1.12) can be rewritten as follows:

$$S_R(x_1) = T_{obs} \cdot sinc\left(T_{obs}\frac{2\Omega}{\lambda}\left(x_1 - x_{1_P}\right)\right) \tag{1.13}$$

It follows that the cross-range resolution is:

$$\delta_{x_1} = \frac{\lambda}{2\Omega T_{obs}} \tag{1.14}$$

For a uniform target rotation, $\Omega T_{obs} \simeq \Delta\vartheta$ where $\Delta\vartheta$ represents the total target aspect angle variation experienced by the radar within the observation time.

From Equation (1.14) it is clear that the higher T_{obs}, the larger $\Delta\vartheta$, the finer the cross-range resolution, δ_{x_1}.

At the end, without any modification of the antenna element, a finer cross-range resolution can be obtained by exploiting the relative motion between the radar and the target.

1.2 ISAR VERSUS SAR

As mentioned in the previous section, real aperture antenna or real antenna arrays do not provide a viable solution for a radar imaging system. To deal with this issue, the SAR concept has been formulated at the beginning of the 50s by Carl Wiley [11, 2] and it is based on the use of an antenna that moves along a given trajectory, therefore providing the means of forming a virtual array in a given time interval. This concept is depicted in Figure 1.3, where a synthetic aperture array formation is compared with a real array. As the SAR image formation is not instantaneous, the equivalence between a real

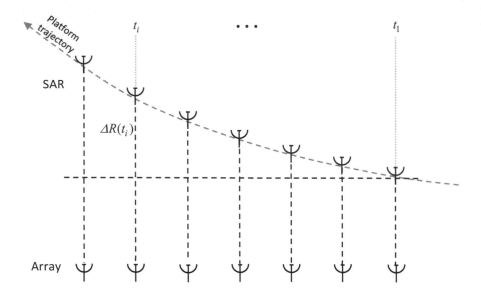

Figure 1.3 SAR system concept

antenna array and the synthetic one holds only if the illuminated area can be considered static during the synthetic aperture formation.

At this point, it is worth pointing out that what is really important to get a radar image of a target is the "relative" motion that exists between the radar and the target itself, as such motion is not necessarily produced by the platform that carries the radar.

This concept has opened the doors to ISAR systems in which the radar is assumed to be stationary while the target is moving [9, 3, 6]. A pictorial representation of an ISAR system is shown in Figure 1.4. As can be seen, as the target moves the aspect angle ϑ changes thus permitting to achieve a desired cross-range resolution. Since the cross-range profile of a target is a function of the radial velocity of each scatterer with respect to the radar, it does not matter what is moving, it could be either the radar, or the target, or even both. Then, according to this point, the difference between a SAR system and an ISAR system seems to be only in the definition of the reference system, and specifically on the point on which the reference system is centered. In fact, by placing the reference system on the target, the SAR geometry is enabled as the target is assumed stationary, conversely by placing the reference system on the radar the ISAR geometry is enabled as the radar is stationary.

Indeed, a subtle but significant difference between the two systems exists. This difference relies on the constraint that must be met to obtain a high resolution SAR or ISAR image. In fact, to get a radar image, the echoes received by the radar at different aspect angles must be coherently processed

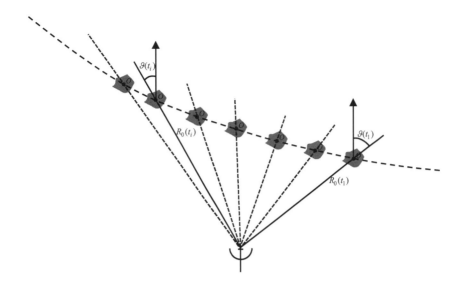

Figure 1.4 ISAR system concept

and then the path differences between each antenna and the target must be compensated. Because of the platform motion, in fact, the distance between the antenna and the target may change over time, thus producing a received signal phase term that changes over time and, in turn, prevents a coherent pulse integration. With reference to Figure 1.3, for example, to coherently process the echoes received by the radar at each time t_i, and then emulate a real antenna array, the phase difference between the signals received by antenna 1 and antenna i must be compensated. When the $straight - iso - range$ approximation holds, which means that the wavefront of the incident e.m. field can be considered planar, and by assuming a target at the antenna broad-side direction, the phase difference between signals received by the antenna at time t_i and at time t_1 depends on ΔR_i, as depicted in Figure 1.3.

The same concept applies to ISAR systems. From Figure 1.4 it is clear that the radar-target distance at each time, $R_0(t)$, changes. To coherently process the echoes received within the observation time, such changes must be compensated.

To do that, the relative motion of the target with respect to the radar must be a priori known. Then, the main difference between SAR and ISAR systems follows. In SAR systems, the radar is mounted on a moving platform and the motion of the radar with respect to the ground is known with adequate precision. Conversely in ISAR systems, the target is usually non-cooperative and the relative motion between the radar and the target is unknown and it must be estimated. The target motion estimation is usually referred to in literature as the *autofocusing* algorithm [3], since it is performed by only using the acquired data without any a priori information.

However, it must be said that both SAR and ISAR processing have drawbacks. SAR processing compensates for the platform motion with respect to a stationary and fixed point on the ground, which is called *focusing point*. As a consequence, a SAR system is able to form a well-focused radar image of all the objects that are static within the illuminated area nearby the *focusing point*. Since SAR processing cannot account for moving targets in the scene, these targets will appear defocused and displaced in the SAR image because of their residual motion with respect to the *focusing point*.

Conversely, the *autofocusing* technique used in ISAR systems accounts for both the radar motion and target motion, as it estimates the relative motion between them. As a drawback, ISAR systems can only process a single target at a time, since different targets have different motions with respect to the radar.

To overcome these limitations, both SAR and ISAR techniques may be jointly used. The idea is to exploit SAR systems to illuminate a large area of interest in which stationary and moving targets may be present. After the SAR image has been formed, ISAR techniques can be used to focus the moving targets in the scene. At the end, well-formed radar images of both the stationary background and the moving targets can be formed which can be used for further applications, such as ATC and ATR.

1.3 RECEIVED SIGNAL MODEL

The objective of this section is to give a definition of the received signal model that can be used for both SAR and ISAR mathematical formulation. For the sake of simplicity, in this section only a monostatic geometry is considered. However, as it will be shown in the following chapters, the same notation will be used for bistatic and passive ISAR systems.

Let the SAR and ISAR geometry be defined in Figure 1.5 and Figure 1.6, respectively.

For both the geometries three different Cartesian reference systems can be identified, namely $T_\xi(\xi_1, \xi_2, \xi_3)$, $T_x(x_1, x_2, x_3)$, and $T_y(y_1, y_2, y_3)$. These reference systems have been defined so they have the same definition in both the ISAR and the SAR geometry. $T_\xi(\xi_1, \xi_2, \xi_3)$ is the absolute reference system. The origin of T_ξ for SAR is on the ground whereas for ISAR it is on the target. $T_x(x_1, x_2, x_3)$ is instead a reference system embedded on the target and whose origin is on the target, namely on the center of the SAR scene for SAR and on a reference point O on the target for ISAR. In both cases, the origin of the T_x reference system, namely the point O, is also called *focusing point*. As the reference system is embedded on the target, in the SAR geometry, it is stationary over time as the target is stationary, while for ISAR as the target is moving the reference system T_x moves accordingly, so that each point on the target is represented by stationary coordinates in T_x.

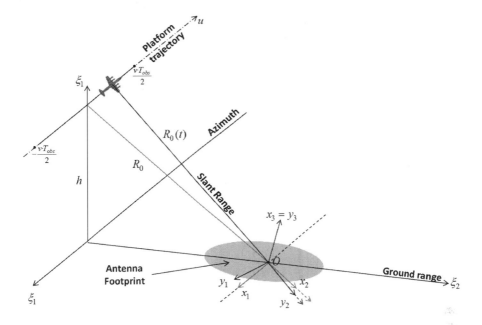

Figure 1.5 SAR geometry and references systems

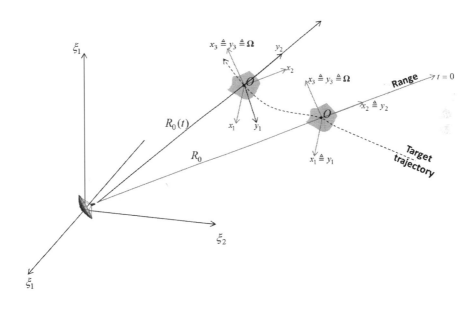

Figure 1.6 ISAR geometry and references systems

In the SAR geometry, the radar moves along the u axis, which is parallel to ξ_1 at a height h from the ground. The reference system T_x is chosen so that the x_2 axis is aligned with radar range direction at the central time $t = 0$ of the observation time, T_{obs}, x_1 is parallel to ξ_1 and x_3 is perpendicular to the plane (x_1, x_2).

For ISAR case, the x_2 axis coincides with the radar radial direction at $t = 0$, x_3 is in the same direction of $\boldsymbol{\Omega}$ (i.e., the so-called *target effective rotation vector* which will be introduced and discussed in Chapter 3) and x_1 is orthogonal to the plane (x_2, x_3).

$T_y (y_1, y_2, y_3)$ is instead a time varying reference system, in other words the coordinates of the three unitary vectors defining the reference system, changes over time with respect to the reference system T_x.

In the SAR case, y_2 is aligned always with the radar radial direction at any time t, y_3 coincides with x_3 and y_1 is orthogonal to the plane (y_2, y_3). For the ISAR case, the y_2 axis is still coincident with the radar radial direction at any time t, the y_3 axis coincides with the x_3 axis and y_1 is perpendicular to the plane (y_2, y_3).

In both the SAR and the ISAR case, the IPP coincides with the plane (x_1, x_2), where the x_2 axis represents the range (or slant-range for SAR) direction, while the x_1 axis represents the cross-range direction. In both cases, therefore, both the x_3 and the y_3 axis are always perpendicular to the IPP.

The definition of the two reference systems, T_x and T_y, allows for a simple identification of the aspect angle variation, which is responsible for the synthetic aperture formation. The aspect angle ϑ is then defined as the angle between x_1 and y_1 or, equivalently, between x_2 and y_2 and it lies in the IPP.

R_0 represents the distance between the radar and the focusing point 0 for $t = 0$, while for a generic time instant the same distance is identified by $R_0(t)$.

Let $s_T(t)$ be a pulse of temporal duration T_I, where usually $T_I \ll T_R$ and T_R is the pulse repetition time (PRT). Let $f'(\mathbf{x})$ be the target reflectivity function defined in the volume V.

Then, the signal received by the radar can be defined as follows:

$$s_R(t) = C \int_V f'(\mathbf{x}) s_T(t - \tau(\mathbf{x}, t)) \, d\mathbf{x} \qquad (1.15)$$

where C is a complex amplitude depending on the radar equation, V is the volume in the 3D spatial domain where the target reflectivity function, $f'(\mathbf{x})$, is defined, $\mathbf{x}$ is the vector that locates the position of an arbitrary point on the target, $\tau(\mathbf{x}, t) = 2R(\mathbf{x}, t)/c$ is the round trip delay time of a point on the target, $R(\mathbf{x}, t)$ is the distance between a point on the target and the radar and c is the light speed in a vacuum.

It is worth pointing out that Equation (1.15) is however an approximation of the received signal, which relies on the superposition principle and neglects the interactions between scatterers.

In typical operating conditions (usually verified for pulse radars), the round trip delay (the time for the e.m. wave to propagate from the radar to the illuminated area and back) is short enough to neglect the relative motion between the radar and the target. This assumption is conventionally called *stop&go* assumption, which implies that the transmission of $s_T(t)$ and the reception of its echo occurs instantaneously. This assumption implies that $R(\mathbf{x}, t)$ can be approximated as $R(\mathbf{x}, n)$, where n is the index that identifies the nth transmitted pulse namely at $t = nT_R$

In other words, if during the whole observation time T_{obs} the radar transmits N pulses (that is, $T_{obs} = NT_R$), the index n identifies the discrete "slow-time" variable.

Then, the signal in a two-dimensional domain, namely the "fast-time" (t) and the "slow-time" (n) domain can be defined as follows:

$$s_R(t, n) = K \cdot rect\left(\frac{n}{N}\right) \int_V f'(\mathbf{x}) s_T(t - \tau(\mathbf{x}, n)) \, d\mathbf{x} \qquad (1.16)$$

where

$$rect\left(\frac{n}{N}\right) \triangleq u\left(n + \frac{N}{2}\right) - u\left(n - \frac{N}{2}\right) = \begin{cases} 1 & -\frac{N}{2} \leq n \leq \frac{N}{2} - 1 \\ 0 & otherwise \end{cases} \qquad (1.17)$$

and

$$u(n) = \begin{cases} 1 & n \geq 0 \\ 0 & otherwise \end{cases} \qquad (1.18)$$

The signal at the output of the matched filter that operates over the *fast-time* variable, t, can be written as follows:

$$s(t, n) = K \cdot rect\left(\frac{n}{N}\right) \int_V f'(\mathbf{x}) C_{s_T}(t - \tau(\mathbf{x}, n)) \, d\mathbf{x} \qquad (1.19)$$

where $C_{s_T}(t)$ is the autocorrelation function of the transmitted signal $s_T(t)$.

Finally by Fourier transforming the signal $s(t, n)$ with respect to the "fast-time" variable, the signal in the frequency/slow-time domain is obtained as follows:

$$S(f, n) = K \cdot rect\left(\frac{n}{N}\right) \cdot |S_T(f)|^2 \int_V f'(\mathbf{x}) \cdot e^{-j2\pi f \tau(\mathbf{x}, n)} \, d\mathbf{x} \qquad (1.20)$$

The received signal formulation as defined in this chapter will be used in the following chapters to go through the SAR and ISAR formulation. Moreover, for the ISAR system both the case of passive radar and bistatic radar will be addressed too.

1.4 CONCLUSIONS

The main radar imaging principles have been introduced in this chapter. First, a definition of the radar image was given. More specifically, a radar image has been defined as an estimate of the spatial distribution of the target reflectivity function mapped onto a 2D plane, arbitrarily oriented in the 3D space, namely the IPP. As well as other type of images, radar images are usually characterized by some quality indexes, among which the geometrical resolutions play a fundamental role, since the finer the spatial resolutions the more detailed the target image. Because of that, a substantial part of the chapter concerns the definition of the radar image spatial resolutions. The radar image spatial resolution depends on the system and, more specifically, it is determined by the system PSF. As it will be shown in the following chapter, for SAR and ISAR systems the PSF can be approximated as a two-dimensional $sinc$-like function whose $-3\ dB$ width along range and cross-range directions corresponds to the range and cross-range resolutions.

Finally, the SAR and ISAR concepts have been introduced and the main differences among them highlighted. More specifically, both SAR and ISAR techniques aim at reconstructing the two-dimensional target reflectivity function mapped onto the IPP by means of some inverse methods. Details of the image reconstruction techniques will be given in Chapters 2 and 3.

REFERENCES

1. W.M. Boerner and H. Überall. *Radar Target Imaging*. Springer Series on Wave Phenomena. Springer, Berlin, Heidelberg, 2011.
2. J.C. Curlander and R.N. McDonough. *Synthetic Aperture Radar: Systems and Signal Processing*. Wiley Series in Remote Sensing and Image Processing. Wiley, 1991.
3. M. Martorella. Introduction to inverse Synthetic Aperture Radar. In S. Theodoridis and R. Chellappa, editors, *Academic Press Library in Signal Processing: Volume 2: Communications and Radar Signal Processing*. Elsevier Science, 2013.
4. D.L. Mensa. *High Resolution Radar Cross-Section Imaging*. Radar Library. Artech House, 1991.
5. C. Oliver and S. Quegan. *Understanding Synthetic Aperture Radar Images*. SciTech Radar and Defense Series. SciTech, 2004.
6. C. Ozdemir. *Inverse Synthetic Aperture Radar Imaging with MATLAB Algorithms*. Wiley Series in Microwave and Optical Engineering. Wiley, 2012.
7. Merrill Ivan Skolnik. *Radar Handbook*. Electronic Engineering Series. McGraw-Hill, 1990.
8. M. Soumekh. *Synthetic Aperture Radar Signal Processing with MATLAB Algorithms*. Wiley-Interscience publication. Wiley, 1999.
9. Jack L. Walker. Range-doppler imaging of rotating objects. *Aerospace and Electronic Systems, IEEE Transactions on*, AES-16(1):23–52, Jan. 1980.
10. D.R. Wehner. *High Resolution Radar*. Artech House Radar Library. Artech House, 1987.

11. C. A. Wiley. Pulse doppler radar methods and apparatus. *United States Patent, No. 3, 196, 436, Filed*, August 1965.

2 SAR Processing

F. Berizzi, S. Tomei, and A. Bacci

CONTENTS

2.1   SAR IMAGING INTRODUCTION

This chapter aims at providing an overview of synthetic aperture radar (SAR) principles and algorithms. The reader who is interested in having further details on SAR is invited to consult [18], [10], [11], [4], [5], [8], [6]. SAR is a signal processing technique applied to coherent radars for remote sensing applications with the capability of effectively imaging areas in all-weather/all-day conditions. In the last few decades, SAR systems have been widely employed for different applications, such as 2D/3D mapping of the Earth's surface, climate change research, environmental monitoring and security-related applications [14, 9, 1].

SAR systems can be either spaceborne or airborne and for both cases the typical side-looking geometry is depicted in Figure 2.1. As the platform moves along its trajectory, the radar illuminates the scene and collects the backscattered energy. The coherent processing of all the echoes from the illuminated scene allows for the synthesis of a long array antenna which is longer than the aperture of the real antenna mounted on the platform. The coherent processing of the received echoes results in a high resolution image of the observed scene. In particular, the SAR image is the 2D reconstruction of the scene electromagnetic reflectivity. As for optical images, one of the main concerns

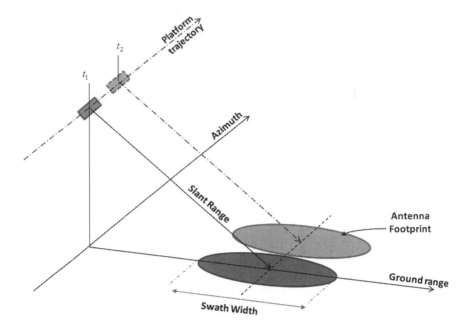

Figure 2.1 SAR geometry

about radar images is to understand how high resolution images can be obtained. In radar images, resolution is defined as the minimum spatial distance for which two point scatterers can be distinguished by the radar along the slant range, x_2, and the azimuth, x_1 with the origin of the system of reference in the center of the scene. The slant range dimension coincides with the line of sight (LoS) of the radar while the azimuth dimension is perpendicular to the slant range dimension and parallel to the ground. The range resolution in conventional imaging radar is inversely proportional to the bandwidth, B, of the transmitted signal

$$\delta_{x_2} = \frac{c}{2B} \tag{2.1}$$

As a consequence, the wider the transmitted bandwidth, the finer the slant range resolution. On the other hand, the resolution in the azimuth direction (or cross-range direction) is inversely proportional to the variation of the aspect angle and, hence, to the synthetic array aperture. In fact, the idea behind SAR is to synthesize the effect of a large physical aperture antenna, the realization of which is impractical because of the size that should have to get a fine enough cross-range resolution, by exploiting the platform motion. In this way, the effects of a long physical antenna are synthetically created in post-processing by coherently processing the echoes received during the syn-

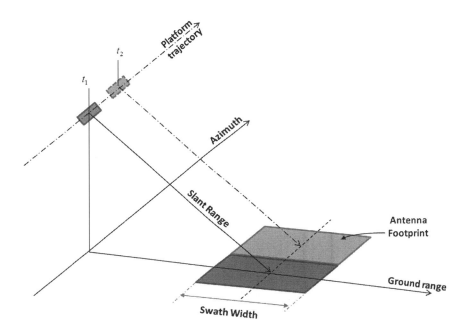

Figure 2.2   SAR stripmap acquisition mode

thetic aperture. The cross-range resolution is inversely related to the synthetic antenna size or array size along the same dimension, L,

$$\delta_{x_1} \propto \frac{1}{L} \tag{2.2}$$

$L = vT_{obs}$ is the *synthetic aperture* and coincides with the platform flight path during the observation time, where v is the platform velocity and T_{obs} the observation time. According to the way the scene is illuminated, a SAR system can be classified as stripmap, spotlight or scansar. As the name suggests, in stripmap SAR the radar illuminates a different spot of the surface being imaged at each instant of time, so that a continuous strip is imaged, as shown in Figure 2.2. On the other hand, a spotlight SAR system illuminates a single spot as the radar platform moves, by steering the antenna beam towards the same area, as shown in Figure 2.3. As a consequence, the same area is illuminated for a longer observation period resulting in a finer resolution. As can be easily understood, the swath width depends on the antenna aperture in azimuth and so it is inherently limited. By the way, in order to increase the swath width, the scansar operational mode can be exploited [11]. In this case, the antenna beam is steered to different elevation angles as the radar platform moves along the flight path, i.e., the azimuth direction, as depicted

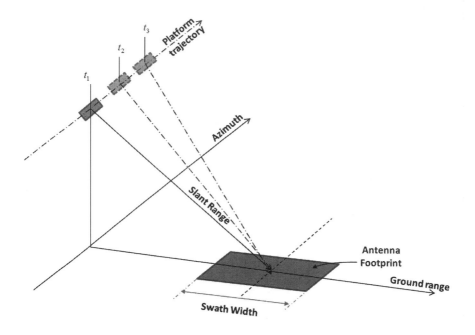

Figure 2.3 SAR spotlight acquisition mode

in Figure 2.4. In this way, by properly setting the elevation offset as the radar moves, a continuous strip can be imaged.

2.2 SAR SIGNAL MODELING

The objective of a SAR system is to form a high resolution image of the illuminated area. In order to describe the various algorithms that can be used to form a focused image from SAR data, let us refer to Figure 2.5 and start from the output of the matched filter at the receiver chain (Equation (1.20))

$$
\begin{aligned}
S\left(f,n\right) &= \int_{V} W\left(f,n\right) f'\left(\mathbf{x}\right) e^{-j2\pi f\tau(\mathbf{x},n)} d\mathbf{x} \\
&= \iint_{x_1 x_2} W\left(f,n\right) f\left(x_1,x_2\right) e^{-j2\pi f\tau(x_1,x_2,n)} dx_1 x_2
\end{aligned}
\tag{2.3}
$$

where $\tau\left(x_1,x_2,n\right)$ is the round-trip delay and $f\left(x_1,x_2\right) = \int_{x_3} f'\left(\mathbf{x}\right) dx_3$ is the projection of the 3D target reflectivity onto the azimuth/range domain. $W\left(f,n\right)$ is the signal support in the frequency/discrete slow time domain, which can be generally rewritten for monostatic radars as

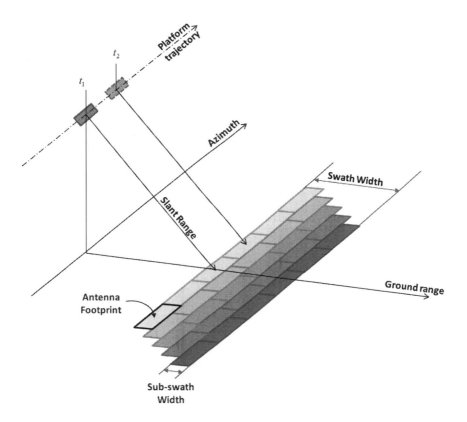

Figure 2.4 SAR scansar acquisition mode

$$W\left(f,n\right) = W_F\left(f\right)W_T\left(n\right) \tag{2.4}$$

where $W_T\left(n\right)$ and $W_F\left(f\right)$ are the signal support in the slow time and frequency domain, respectively. In particular, at the output of the matched filter (MF) the signal support in the frequency domain is given by

$$W_F\left(f\right) = \left|S_T\left(f\right)\right|^2 = \text{rect}\left(\frac{f - f_0}{B}\right) \tag{2.5}$$

where f_0 is the carrier frequency and B is the signal bandwidth. The $rect\left(\cdot\right)$ denotes the continuous rect function. On the other hand, the signal support in the slow time domain depends on the SAR operational mode. In particular, for the stripmap geometry, each scatterer, denoted by k, is illuminated for the same time $T_{obs} = NT_R$ and the support in the slow time domain can be written as

$$W_T\left(n, x_1\right) = \text{rect}\left(\frac{n - n_{0_k}}{N}\right) \tag{2.6}$$

where $n_{0_k} = \frac{x_{1_k}}{vT_R}$, $N = \frac{T_{obs}}{T_R}$ denotes the number of sweeps, v is the radar platform speed and x_{1_k} is the cross-range coordinate of the kth scatterer. In this case, $rect\left(\cdot\right)$ denotes the discrete rect function, given by $rect(\frac{n}{N}) = u\left(n\right) - u\left(n - N\right)$, where $u\left(\cdot\right)$ is the unit step function. It is worth pointing out that the time interval in which the scene is observed depends on the antenna footprint size along the azimuth direction in stripmap systems, while in spotlight systems the radar illuminates the same scene for the whole observation time set during the design of the radar, so that $n_{0_k} = 0$ and

$$W_T\left(n, x_1\right) = W_T\left(n\right) = \text{rect}\left(\frac{n}{N}\right) \tag{2.7}$$

From Figure 2.5, the distance between the radar and a generic point-like scatterer, P, located in (x_{1_k}, x_{2_k}) at the generic time n can be calculated as

$$R\left(n\right) = \sqrt{\left(R_0 + x_{2_k}\right)^2 + \left(vnT_R - x_{1_k}\right)^2} \tag{2.8}$$

which is equivalent to

$$R\left(n\right) = \sqrt{\left[R_0\left(n\right) + y_{2_k}\left(n\right)\right]^2 + y_{1_k}^2\left(n\right)} \tag{2.9}$$

where $y_{1_k}\left(n\right)$ and $y_{2_k}\left(n\right)$ are the coordinates of the scatterer point along the (y_1, y_2) reference system, in which y_2 is directed along the slant range direction of the origin O of the (x_1, x_2) reference system, whereas y_1 is orthogonal to y_2. Under the *straight iso-range approximation*, i.e., $\frac{y_{1_k}\left(n\right)}{R_0\left(n\right)} \ll 1$, the radar-target distance can be approximated as follows:

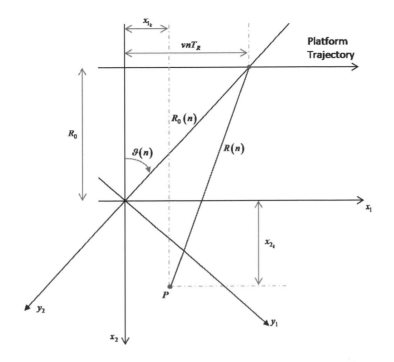

Figure 2.5 SAR geometry in the slant range/azimuth plane

$$R(n) \simeq R_0(n) + y_{2_k}(n)$$
$$= \sqrt{R_0^2 + (vnT_R)^2} + y_{2_k}(n) \tag{2.10}$$
$$= \sqrt{R_0^2 + (vnT_R)^2} + x_{1_k} \sin\vartheta(n) + x_{2_k} \cos\vartheta(n)$$

where $tan(\vartheta(n)) = \frac{vnT_R}{R_0}$ and $\vartheta(n)$ is the aspect angle.
By substituting Equation (2.10) into Equation (2.3) the signal at output of the matched filter becomes

$$S(f,n) = W_F(f) e^{-j\frac{4\pi f}{c}R_0(n)}$$
$$\cdot \iint\limits_{x_1 x_2} W_T(n, x_1) f(x_1, x_2) e^{-j\frac{4\pi f}{c}[x_1 \sin(\vartheta(n)) + x_2 \cos(\vartheta(n))]} dx_1 dx_2$$

$$(2.11)$$

The phase term outside the integral is responsible for the range migration and is due to the distance between the scene center, O, and the platform carrying the radar sensor. Such phase term must be properly compensated to avoid any image blurring and artifacts. The compensation of this phase term is usually denoted by motion compensation step.

At the output of the matched filter, the received signal after motion compensation becomes

$$S_C(f,n) = W_F(f)$$
$$\cdot \iint_{x_1 x_2} W_T(n, x_1) f(x_1, x_2) e^{-j\frac{4\pi f}{c}[x_1 \sin(\vartheta(n)) + x_2 \cos(\vartheta(n))]} dx_1 dx_2$$

$$(2.12)$$

2.3 SAR IMAGE FORMATION TECHNIQUES

The following sections are dedicated to the description of some of the most used SAR image formation algorithms.

2.3.1 RANGE DOPPLER ALGORITHM

The signal at the output of the MF in Equation (2.12) can be rewritten as a function of the spatial frequencies, (X_1, X_2), which are defined as follows:

$$\begin{cases} X_1(f,n) = \frac{2f}{c} \sin(\vartheta(n)) \\ X_2(f,n) = \frac{2f}{c} \cos(\vartheta(n)) \end{cases}$$

$$(2.13)$$

According to Equation (2.13), the signal support in the spatial frequency domain is defined by a sector of annulus as represented in Figure 2.6.

When the aspect angle variation within the observation time, defined as $\Delta\vartheta$, is small, i.e., $|\Delta\vartheta| \ll 1$, then the following approximation holds:

$$\begin{cases} X_1(f,n) \simeq \frac{2f}{c}(\vartheta(n)) = \frac{2f_0 vn T_R}{cR_0} \\ X_2(f,n) \simeq \frac{2f}{c} \end{cases}$$

$$(2.14)$$

and the spatial frequency domain of the signal can be approximated to a rectangular domain, as shown in Figure 2.7, since $X_1(f,n)$ and $X_2(f,n)$ are two separable variables [20].

According to the approximation in Equation (2.14), Equation (2.12) can be simplified as follows:

$$S_C(f,n) = W_F(f) \iint_{x_1 x_2} W_T(n, x_1) f(x_1, x_2) e^{-j2\pi\left[\frac{2f_0 vn T_R}{R_0 c} x_1 + \frac{2f}{c} x_2\right]} dx_1 dx_2$$
$$= W_F(f) F_W(f,n)$$

$$(2.15)$$

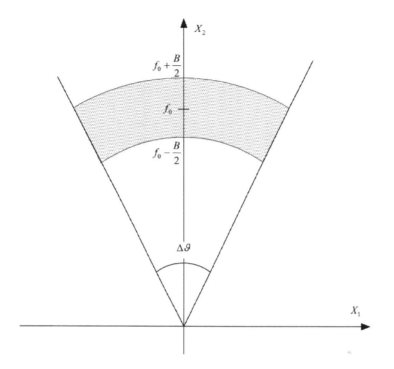

Figure 2.6 Spatial frequency domain

where

$$F_W(f,n) = \iint\limits_{x_1 x_2} W_T(n, x_1) f(x_1, x_2) e^{-j2\pi \left[\frac{2f_0 vnT_R}{R_0 c} x_1 + \frac{2f}{c} x_2\right]} dx_1 dx_2 \quad (2.16)$$

As can be easily noticed from Equation (2.15), the relationship between the motion compensated signal, $S_C(f,n)$, and the target reflectivity function, $f(x_1, x_2)$ weighted for the signal support in the slow-time domain is given by a 2D inverse Fourier transform (IFT)

$$\begin{aligned} I_{RD}(\tau, \nu) &= \text{2D-IFT}\left[S_C(f,n)\right] \\ &= w_F(\tau) \otimes_\tau f_W(\tau, \nu) \end{aligned} \quad (2.17)$$

where $w_F(\tau) = IFT[W_F(f)]$, $\otimes_\tau$ denotes the convolution with respect to the variable τ, $f_W(\tau, \nu) = IFT[F_W(f,n)]$ and (τ, ν) represents the delay time and the Doppler frequency pair. In particular,

$$\begin{cases} \tau = \frac{2x_2}{c} \\ \nu = \frac{2f_0 v x_1}{cR_0} \end{cases} \quad (2.18)$$

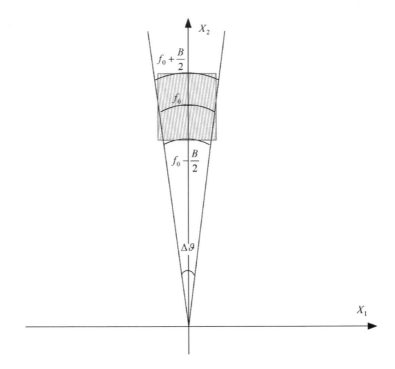

Figure 2.7 Spatial frequency domain approximation for small aspect angle variation $(|\Delta\vartheta| \ll 1)$

so that

$$F_W(f,n) = \frac{c^2 R_0}{4f_0 v} \iint\limits_{\tau\nu} W_T(n,\nu) f\left(\frac{cR_0}{2f_0 v}\nu, \frac{c}{2}\tau\right) e^{-j2\pi[\tau f + \nu n T_R]} d\tau d\nu \quad (2.19)$$

Since Equation (2.16) represents a Fourier transform of a product between $W_T(n, x_1)$ and $f(x_1, x_2)$, Equation (2.17) can be rewritten as

$$I_{RD}(\tau, \nu) = w_F(\tau) w_T(\nu) \otimes_\tau \otimes_\nu f(\tau, \nu) \quad (2.20)$$

where $\otimes_\nu$ denotes the convolution operator with respect to ν and

$$w_t(\nu) = IFT\left[W_T(n, x_1)\right]\Big|_{\tau = \frac{2x_2}{c}}$$

$$f(\tau, \nu) = f(x_1, x_2)\Big|_{\substack{\tau = \frac{2x_2}{c} \\ \nu = \frac{2f_0 v x_1}{cR_0}}} \quad (2.21)$$

It is worth pointing out that the product $w_F(\tau) w_T(\nu)$ identifies the point spread function (PSF) of the SAR system in the delay time/Doppler domain.

According to Equation (2.6) and Equation (2.7), a distinction between the stripmap case and the spotlight case can be performed. In particular, the PSF for the spotlight case is given by

$$PSF_{spot}\left(\tau,\nu\right) = T_{obs}B\mathrm{sinc}\left(B\tau\right)\mathrm{sinc}\left(T_{obs}\nu\right)e^{-j2\pi\tau f_0} \qquad (2.22)$$

where $sinc\left(x\right)$ denotes the $\frac{\sin(\pi x)}{\pi x}$ function. For the stripmap case, the system PSF is given by

$$PSF_{strip}\left(\tau,\nu\right) = T_{obs}B\mathrm{sinc}\left(B\tau\right)\mathrm{sinc}\left(T_{obs}\nu\right)e^{-j2\pi\tau f_0}e^{-j2\pi\nu n_{0_k}} \qquad (2.23)$$

As can be easily noticed, the only difference between the stripmap case and the spotlight case is a phase term which in turn does not affect the image of the target since it coincides with the modulus of the complex image given by the PSF in the case of a single scatterer. For example, suppose the target consists of K point-like scatterers, each one with its own reflectivity value, σ_k, and coordinates (x_{1_k}, x_{2_k}). In this case, the 2D global reflectivity function[1] can be written as follows:

$$f\left(x_1,x_2\right) = \sum_{k=1}^{K}\sigma_k\delta\left(x_1 - x_{1_k}, x_2 - x_{2_k}\right) \qquad (2.24)$$

where $\delta\left(\cdot,\cdot\right)$ denotes the 2D Dirac delta. Equation (2.24) can be rewritten in the (τ,ν) domain by exploiting the relationships in Equation (2.18) as follows:

$$f\left(\tau,\nu\right) = \sum_{k=1}^{K}\sigma_k\delta\left(\tau - \tau_k, \nu - \nu_k\right) \qquad (2.25)$$

The corresponding SAR image can be rewritten as

$$I_{RD}\left(\tau,\nu\right) =$$
$$= \sum_{k=1}^{K}T_{obs}B\mathrm{sinc}\left(B\left(\tau - \tau_k\right)\right)\mathrm{sinc}\left(T_{obs}\left(\nu - \nu_k\right)\right) \qquad (2.26)$$
$$\cdot\, e^{-j2\pi(\tau-\tau_k)f_0}e^{-j2\pi(\nu-\nu_k)n_{0,k}}$$

Considering the relationships in Equation (2.18), a scaling operation can be easily applied in order to obtain the complex SAR image in the range/azimuth domain. In particular, the scaling relationships are given by

---

[1]Under the assumption that the interactions among the scatterers can be neglected.

$$\begin{cases} x_2 = \frac{c\tau}{2} \\ x_1 = \frac{vcR_0}{2f_0v} \end{cases} \tag{2.27}$$

and the scaled signal is

$$I_{RD}\left(x_1, x_2\right) = \sum_{k=1}^{K} T_{obs} B \mathrm{sinc}\left(\frac{2T_{obs}f_0v}{cR_0}\left(x_1 - x_{1_k}\right)\right) \mathrm{sinc}\left(\frac{2B}{c}\left(x_2 - x_{2_k}\right)\right)$$
$$\cdot e^{-j2\pi \frac{cR_0}{2f_0v}\left(x_1 - x_{1_k}\right)n_{0_k}} e^{-j2\pi \frac{2B}{c}\left(x_2 - x_{2_k}\right)f_0} \tag{2.28}$$

A block diagram that summarizes the steps of the RD algorithm is in Figure 2.8, where the two-dimensional fast Fourier transform (2D-FFT) can be applied by taking into account that the signal $S_C\left(f, n\right)$ is also sampled in the frequency domain.

Figure 2.8 Block diagram of the RD algorithm

The resolution in the range domain, δ_r, and in the azimuth domain, δ_{cr}, can be easily found. They can be defined as the first null of the $sinc\left(\cdot\right)$ functions [3]

$$\begin{cases} \delta_r = \delta_{x_1} = \frac{c}{2B} \\ \delta_{cr} = \delta_{x_2} = \frac{cR_0}{2f_0L_{cr}} \end{cases} \tag{2.29}$$

where $L_{cr} = vT_{obs}$ denotes the synthetic aperture.

2.3.2 POLAR-FORMAT

When the aspect angle variation, $\Delta\vartheta$, in the observation time is not small enough to allow the approximation of the spatial frequency domain to a rectangular domain, a different image formation algorithm must be applied. Under this assumption, the polar format (PF) algorithm can be applied in the $\left(X_1, X_2\right)$ domain [20]. In this case, the signal samples in the $\left(X_1, X_2\right)$ domain are located on a polar grid, as represented in Figure 2.9 in light grey solid line. This means that the signal samples are not equally spaced, so that in order to obtain the SAR image a discrete Fourier transform (DFT) should be applied. By the way, on one hand the DFT can deal with not equally spaced samples and on the other hand it requires high computational time. In order to reduce the processing time, the DFT can be replaced by the FFT algorithm, but a

2D interpolation (2D-INT) step is mandatory to map the signal samples in an equally spaced rectangular grid, as shown in Figure 2.9 with black solid lines.

The block diagram of the PF process is in Figure 2.10, which shows the algorithm when the 2D interpolation is applied.

Following the block diagram in Figure 2.10, the motion compensated signal at the input of the 2D interpolation process is given by

$$S_C(f, n) = W_F(f) \iint\limits_{x_1 x_2} W_T(n, x_1) f(x_1, x_2) e^{-j2\pi[X_1 x_1 + X_2 x_2]} dx_1 dx_2 \quad (2.30)$$

where (X_1, X_2) are the spatial frequencies defined in Equation (2.13). Let $S_{\text{2D-INT}}(X_1, X_2)$ be the signal after the 2D interpolation step.

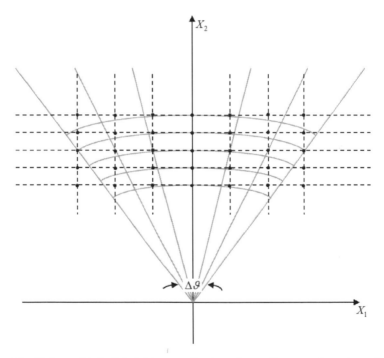

Figure 2.9 Polar grid (solid light gray line) and equally spaced rectangular grid (dotted black line) after 2D interpolation step

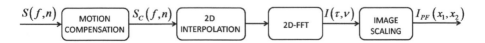

Figure 2.10 Block diagram of the PF algorithm

At this stage, the SAR image can be obtained as

$$I_{PF}(x_1, x_2) = \text{2D-FFT}\left[S_{\text{2D-INT}}(X_1, X_2)\right] \tag{2.31}$$

It should be noted that the PF technique is specially suited for spotlight SAR applications, in which the same scene is observed for a long observation time, usually corresponding to a large variation of the aspect angle [15]. It is worth remarking that the PF algorithm can be applied in SAR since $\vartheta(n)$ is well known in the system geometry and the polar grid in the spatial frequency domain can be exactly determined.

2.3.3 OMEGA-K

This SAR image formation algorithm is based on the processing of the signal at the output of the matched filter before the motion compensation step.

According to the geometry in Figure 2.5, the signal at the output of the matched filter (see Equation (2.3)) can be easily rewritten as

$$S(f, u_n) = W_F(f) \iint_{x_1 x_2} W_T(u_n)\, f(x_1, x_2)\, e^{-j\frac{4\pi f}{c}\sqrt{(R_0+x_2)^2+(u_n-x_1)^2}}\, dx_1 dx_2$$

$$\tag{2.32}$$

where $u_n = vnT_R$, represents the platform cross-range coordinate.

The Fourier transform with respect to u_n can be performed by exploiting the *principle of stationary phase* [19], so that

$$S(f, k_u) =$$
$$= W_F(f) \iint_{x_1 x_2} W_n(k_u, x_1, x_2)\, f(x_1, x_2)\, e^{-j\sqrt{4\left(\frac{2\pi f}{c}\right)^2-k_u^2}\cdot(x_2+R_0)-jk_u x_1}\, dx_1 dx_2$$

$$\tag{2.33}$$

where $W_F(f)$ is defined in Equation (2.5), k_u is the transformed variable with respect to u_n and $W_T(k_u)$ denotes the signal support in the k_u domain.

The motion compensation is performed via multiplication with $S_0^*(f, k_u)$, where

$$S_0(f, k_u) = e^{-j\sqrt{4\left(\frac{2\pi f}{c}\right)^2-k_u^2}R_0}\, W_F(f)\, W_{T0}(k_u) \tag{2.34}$$

which corresponds to the signal return from a point-like scatterer located in $(x_{1_k} = 0, x_{2_k} = 0)$. In particular, $W_{T0}(k_u) = W_T(k_u, x_1 = 0, x_2 = 0)$ is the signal support for such a scatterer.

The motion compensated signal can be rewritten as

$$S_C(f, k_u) =$$
$$= |W_F(f)|^2 W_T(k_u, x_1, x_2) \, W_{T0}(k_u) \iint\limits_{x_1 x_2} f(x_1, x_2) \, e^{-j\sqrt{4\left(\frac{2\pi f}{c}\right)^2 - k_u^2} \, x_2 - j k_u x_1} \, dx_1 \, dx_2$$

$$(2.35)$$

If the illuminated region is small with respect to the radar-ground distance, i.e., $|x_1| \ll R_0$ and $|x_2| \ll R_0$, then the following approximation holds:

$$W_T(k_u, x_1, x_2) \, W_{T0}(k_u) \simeq W_T(k_u, x_1, x_2) \qquad (2.36)$$

and the signal after motion compensation becomes

$$S_C(f, k_u) =$$
$$|W_F(f)|^2 \, W_T(k_u, x_1, x_2) \iint\limits_{x_1 x_2} f(x_1, x_2) \, e^{-j\sqrt{4\left(\frac{2\pi f}{c}\right)^2 - k_u^2} \cdot x_2 - j k_u x_1} \, dx_1 \, dx_2$$

$$(2.37)$$

In order to understand the signal spectral occupancy in the k_u domain, it is useful to analyze the signal instantaneous frequency, F_{k_u}, which is defined as

$$
\begin{aligned}
F_{k_u} &= \frac{\partial}{\partial u_n} \left[-\frac{2\pi f}{c} \left(\sqrt{(R_0 + x_2)^2 + (u_n - x_1)^2} \right) \right] \\
&= -\frac{2\pi f}{c} \frac{u_n - x_1}{\sqrt{(R_0 + x_2)^2 + (u_n - x_1)^2}} \\
&= -\frac{2\pi f}{c} \sin\left(\vartheta\left(u_n, x_1, x_2\right)\right)
\end{aligned}
\qquad (2.38)
$$

where $\vartheta(u_n, x_1, x_2)$ is the aspect angle of a scatterer located in (x_1, x_2).

By recalling that $u_n = v n T_R$ represents the radar platform position along the trajectory at the time $n T_R$, it is easy to see that u_n ranges from $-\frac{L}{2}$ and $+\frac{L}{2}$. When the antenna patch size is much smaller than the distance R_0, the aspect angle $\vartheta(u_n, x_1, x_2)$ can be approximated by $\vartheta(u_n, x_1, x_2) \approx \vartheta(u_n, 0, 0) = \vartheta(u_n)$. In this case, the signal support in the Doppler angular domain (k_u) occupies a Doppler bandwidth defined by

$$B_{k_u} = \left[-\frac{2\pi f}{c} \sin\left(\vartheta\left(\frac{L}{2}\right)\right), -\frac{2\pi f}{c} \sin\left(\vartheta\left(-\frac{L}{2}\right)\right) \right] \qquad (2.39)$$

As a consequence, the signal support in the angular Doppler domain can be approximated as follows:

$$W_T\left(k_u, x_1, x_2\right) \simeq \begin{cases} 1 & k_u \in B_{k_u} \\ 0 & elsewhere \end{cases} \tag{2.40}$$

A representation of the signal support in the (f, k_u) domain is given in Figure 2.11.

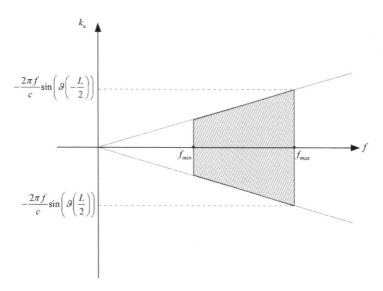

Figure 2.11 Signal support representation in the (f, k_u) domain where $f_{min} = f_0 - \frac{B}{2}$ and $f_{max} = f_0 + \frac{B}{2}$

By defining the following spatial frequencies

$$\begin{cases} X_1 = k_u \\ X_2 = \sqrt{4\left(\frac{2\pi f}{c}\right)^2 - k_u^2} \end{cases} \tag{2.41}$$

the support of the signal in Equation (2.37) can be rewritten in the spatial frequency domain as

$$W_k\left(X_1, X_2\right) \simeq \begin{cases} 1 & \left(X_1, X_2\right) \in D_k\left(X_1, X_2\right) \\ 0 & elsewhere \end{cases} \tag{2.42}$$

where the the dependency on k denotes that the signal support in the spatial frequency domain depends on the scatterer location. $W_k\left(X_1, X_2\right)$ is represented in Figure 2.12.

By exploiting the spatial frequency definition in Equation (2.41), the signal at the output of the matched filter after motion compensation can be rewritten as

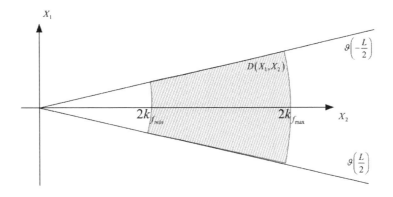

Figure 2.12 Signal support approximation in the (X_1, X_2) domain

$$S_C\left(X_1, X_2\right) = \iint\limits_{x_1, x_2} W_k\left(X_1, X_2\right) f\left(x_1, x_2\right) e^{-j[x_1 X_1 + x_2 X_2]} dx_1 dx_2 \qquad (2.43)$$

By referring to Equation (2.43), it is clear that in the case of a single scatterer with reflectivity σ_k, the motion compensated signal is given by

$$S_C\left(X_1, X_2\right) = \iint\limits_{x_1, x_2} W_k\left(X_1, X_2\right) \sigma_k\left(x_1 - x_{1_k}, x_2 - x_{2k}\right) e^{-j[x_1 X_1 + x_2 X_2]} dx_1 dx_2$$

$$(2.44)$$

The corresponding image, which defines the PSF of the system, is given by the Fourier transform of the motion compensated signal

$$
\begin{aligned}
I_{OK}\left(x_1, x_2\right) &= FT\left[S_C\left(X_1, X_2\right)\right] \\
&= w_k\left(x_1 - x_{1k}, x_2 - x_{2k}\right)
\end{aligned}
\qquad (2.45)
$$

As can be easily noticed, the PSF is spatially variant, which means that each scatterer is imaged with a different impulse response centered on the scatterer location in the (x_1, x_2) domain.

In addition, it should be noted that the samples of the signal in Equation (2.44) are defined in a not-uniform and not-rectangular grid in the (f, k_u) domain, as it can be easily noticed by Figure 2.11. Then, it is clear that a DFT should be applied, which is highly time consuming. In order to reduce the computational load of the algorithm, a 2D interpolation must be performed in order to apply a conventional 2D-FFT. A block diagram of the Omega-K image formation algorithm is shown in Figure 2.13.

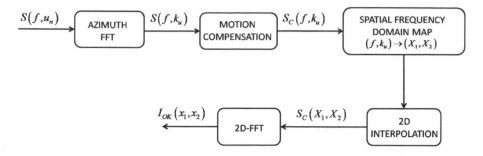

Figure 2.13 Block diagram of the Omega-K image formation algorithm

By observing the spatial frequency domain representation in Figure 2.12, it is easy to understand that for small aspect angle variation, the domain can be approximated with a rectangular domain, as shown in Figure 2.14 where B_{X_1} and B_{X_2} denote the signal bandwidths along two orthogonal directions. In detail, it can be easily demonstrated that

$$
\begin{cases}
B_{X_2} \simeq \frac{4\pi B}{c} \\
B_{X_1} \simeq 2 \left(\frac{2\pi f_0}{c} \right) \left[\vartheta \left(\frac{L}{2} \right) - \vartheta \left(-\frac{L}{2} \right) \right] \approx \frac{4\pi}{\lambda_0} \frac{R_n}{L \cos(\vartheta(0))}
\end{cases}
\tag{2.46}
$$

which lead to the range and cross-range resolution of the SAR image. In particular,

$$
\begin{cases}
\delta_{x_2} = \frac{2\pi}{B_{X_2}} = \frac{c}{2B} \\
\delta_{x_1} = \frac{2\pi}{B_{X_1}} = \frac{\lambda_0}{2} \frac{L \cos(\vartheta(0))}{R_n}
\end{cases}
\tag{2.47}
$$

Let $S_{\text{2D-INT}}(X_1, X_2)$ be the signal at the output of the 2D interpolation step, so that the complex SAR image can be obtained by applying a 2D IFT to the signal in Equation (2.44), through 2D-IFT algorithm

$$
I_{OK}(x_1, x_2) = \text{2D-IFT}\left[S_C(X_1, X_2) \right]
\tag{2.48}
$$

In particular, in the case of a single scatterer target, located at (x_{1_k}, x_{2_k}), the SAR image can be rewritten as

$$
\begin{aligned}
I_{OK}(x_1, x_2) &= w(x_1, x_2) \otimes \otimes \left[\sigma_k \delta\left(x_1 - x_{1_k}, x_2 - x_{2_k} \right) \right] \\
&= w(x_1 - x_{1_k}, x_2 - x_{2_k})
\end{aligned}
\tag{2.49}
$$

where $w(x_1 - x_{1_k}, x_2 - x_{2_k})$ represents the impulsive response of the SAR system. It can be easily noticed that when the support bandwidth in the angular Doppler domain is small enough to allow for the domain approximation shown in Figure 2.14, the impulsive response does not depend on the scatterer location, so that each scatterer is imaged with the same impulsive response.

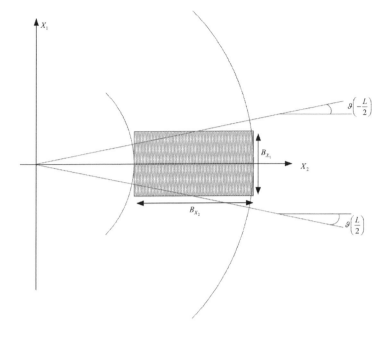

Figure 2.14 Signal support approximation in the (X_1, X_2) domain

2.3.4 BACK-PROJECTION

In this case, no approximation is assumed for the spatial frequency domain and the SAR image is obtained considering the samples collected on the polar grid as shown in Figure 2.9 [7, 16, 2, 20]. It is worth pointing out that the back-projection (BP) algorithm is especially suited for spotlight SAR, in which the signal after motion compensation can be written as

$$S_C\left(f, n\right) = W\left(f, n\right) \iint\limits_{x_1 x_2} f\left(x_1, x_2\right) e^{-j2\pi[x_1 \sin \vartheta(n) + x_2 \cos \vartheta(n)]\Gamma} dx_1 x_2 \quad (2.50)$$

where $\Gamma = \frac{2f}{c}$ and $W\left(f, n\right) = W_F\left(f\right) W_T\left(n\right)$ with $W_T\left(n\right)$ defined in Equation (2.7). The inner integral in Equation (2.50) corresponds to the Fourier transform of the projection of the target reflectivity function, $f\left(x_1, x_2\right)$, at an angle $\vartheta\left(n\right)$. At each aspect angle $\vartheta\left(n\right)$, with $n = -\frac{N}{2}, \cdots, 0, \cdots, \frac{N}{2} - 1$, a different target reflectivity function projection, $p_{\vartheta(n)}\left(\gamma\right)$, is defined as

$$p_{\vartheta(n)}\left(\gamma\right) = \iint\limits_{x_1 x_2} f\left(x_1, x_2\right) e^{-j2\pi[x_1 \sin \vartheta(n) + x_2 \cos \vartheta(n)]\Gamma} dx_1 x_2 \quad (2.51)$$

In particular, considering the Fourier transform of the projection, $P_\vartheta(\Gamma)$, the following equivalence holds:

$$
\begin{aligned}
P_{\vartheta(n)}(\Gamma) &= \int_\gamma p_{\vartheta(n)}(\gamma) e^{-j2\pi\Gamma\gamma} d\gamma \\
&= \int_\gamma \left[\iint_{x_1 x_2} f(x_1, x_2) \delta(x_1 \sin\vartheta(n) + x_2 \cos\vartheta(n) - \gamma) dx_1 dx_2 \right] \times \\
&\quad e^{-j2\pi\Gamma\gamma} d\gamma \\
&= \iint_{x_1 x_2} f(x_1, x_2) e^{-j2\pi\Gamma(x_1 \sin\vartheta(n) + x_2 \cos\vartheta(n))} dx_1 x_2
\end{aligned}
$$

$$(2.52)$$

Equation (2.52) represents the *projection-slice theorem*, for which the Fourier transform of the projection of an object function at an angle $\vartheta(n)$ corresponds to a slice of the Fourier transform of the object function at the same angle.

By exploiting the definition of Γ and the spatial frequencies defined in Equation (2.13), Equation (2.50) can be rewritten in the spatial frequency domain as

$$
S_C(X_1, X_2) = W(X_1, X_2) \Theta(X_1, X_2) \tag{2.53}
$$

where $W(X_1, X_2)$ is the signal support function in the spatial frequency domain (X_1, X_2) defined in Equation(2.13) and where

$$
\Theta(X_1, X_2) = \iint_{x_1 x_2} f(x_1, x_2) e^{-j2\pi\Gamma(x_1 \sin\vartheta(n) + x_2 \cos\vartheta(n))} dx_1 x_2 \tag{2.54}
$$

A comparison between Equation (2.54) and Equation (2.52) shows that the motion compensated signal in the (X_1, X_2) is related to the Fourier transform of the projection of the target reflectivity function. In particular,

$$
P_{\vartheta(n)}(\Gamma) = \Theta(X_1, X_2) \Big|_{\substack{X_1 = \Gamma \sin\vartheta(n) \\ X_2 = \Gamma \cos\vartheta(n)}} \tag{2.55}
$$

For this reason, the complex SAR image can be obtained as

$$
\begin{aligned}
I_{BP}(x_1, x_2) &= \text{2D-IFT}\{S_C(X_1, X_2)\} \\
&= \iint_{X_1 X_2} W(X_1, X_2) \Theta(X_1, X_2) e^{-j2\pi[x_1 X_1 + x_2 X_2]} dX_1 dX_2 \tag{2.56}
\end{aligned}
$$

Since the samples are acquired on a polar grid domain, it is useful to use the polar representation so that the following transformation is applied to Equation (2.56):

$$\begin{cases} \Gamma = \sqrt{X_1^2 + X_2^2} \\ \Phi = \arctan \frac{X_1}{X_2} \end{cases} \tag{2.57}$$

The complex SAR image is obtained in a polar domain (γ, ϕ)

$$
\begin{aligned}
I_{BP}(\gamma, \phi) &= \int_{\Phi=-\frac{\pi}{2}}^{+\frac{\pi}{2}} \int_{\Gamma=0}^{\infty} W(\Gamma, \Phi) |\Gamma| \Theta(\Gamma, \Phi) e^{j2\pi\gamma \cos(\Phi-\phi)} d\Gamma d\Phi \\
&= \int_{\Phi=-\frac{\pi}{2}}^{+\frac{\pi}{2}} \int_{\Gamma=0}^{\infty} W(\Gamma, \Phi) |\Gamma| \{P_{\vartheta(n)}(\Gamma)\}_{\vartheta(n)=\Phi} e^{j2\pi\gamma \cos(\Phi-\phi)} d\Gamma d\Phi \\
&= \int_{\Phi=-\frac{\pi}{2}}^{+\frac{\pi}{2}} \int_{\Gamma=0}^{\infty} W_{BP}(\Gamma, \Phi) \Theta(\Gamma, \Phi) e^{j2\pi\gamma \cos(\Phi-\phi)} d\Gamma d\Phi \\
&= \int_{\Phi=-\frac{\pi}{2}}^{+\frac{\pi}{2}} q_{\vartheta(n)=\Phi}(x_1 \sin(\Phi) + x_2 \cos(\Phi)) d\Phi
\end{aligned}
\tag{2.58}
$$

where $W_{BP}(\Gamma) = W(\Gamma, \Phi) |\Gamma|$.

The second line in Equation (2.58) corresponds to the IFT of the product $W_{BP}(\Gamma, \Phi) \Theta(\Gamma, \Phi)$ evaluated for $\vartheta(n) = \Phi$, and it is usually referred to as the *filtered projection*. $W_{BP}(\Gamma, \Phi)$ is the signal support filtered with a ramp filter given by $|\Gamma| = \left| \frac{2f}{c} \right|$. In order to find the SAR image in the (x_1, x_2) domain, a transformation from the polar domain, (γ, ϕ), to the rectangular domain, (x_1, x_2) is needed. It is worth pointing out that since in practical cases $\vartheta(n)$ is a discrete variable, the SAR image is estimated as follows:

$$I_{BP}(x_1, x_2) \simeq \sum_{n=-\frac{N}{2}}^{+\frac{N}{2}-1} q_{\vartheta(n)}(x_1 \sin(\vartheta(n)) + x_2 \cos(\vartheta(n))) \tag{2.59}$$

In this case, it is obvious that an interpolation step is required to map the data in a discrete grid.

A block diagram showing the steps of the BP algorithm is in Figure 2.15.

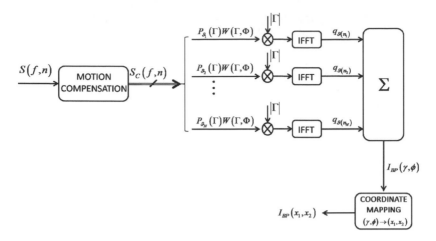

Figure 2.15   Back projection block diagram

2.3.5   CHIRP SCALING

In the previous sections it has been shown that in order to compensate the range cell migration, conventional SAR image formation algorithms exploit a space-variant interpolation, which involves significant computational time and leads to a loss of image quality [12, 17, 13]. The chirp scaling (CS) algorithm, also known as *differential range deramp* (DRD), performs range cell migration correction (RCMC) avoiding the interpolation step, yet leading to an image quality which is equal to or better than range-Doppler-based algorithms.

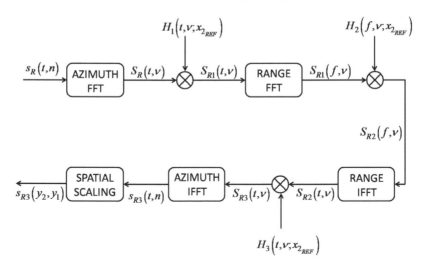

Figure 2.16   Chirp scaling block diagram

The basic idea of the CS algorithm consists of performing an equalization of the range cell migration with respect to a reference cell in the range/Doppler domain, followed by range and pulse compression. Once back in the range-Doppler domain, a residual phase correction is performed. In order to describe the CS algorithm, let us consider the received signal in the fast time/slow time domain, given by

$$s_R\left(t, n; x_{2_k}\right) = C \iint_{x_1 x_2} f\left(x_1, x_2\right) s_T\left(t - \tau\left(x_1, x_2; x_{2_k}\right)\right) dx_1 dx_2 \qquad (2.60)$$

where x_{2_k} denotes the range coordinate of the target scatterer at the radar-target closest approach and $\tau\left(x_1, x_2; x_{2_k}\right)$ denotes the range delay term with respect to the scatterer range at the closest approach to the radar. Assume that the transmitted signal is a chirp of chirp rate μ,

$$s_T\left(t, n\right) = e^{j2\pi\left(f_0 t + \mu t^2\right)} w_P\left(t\right) w_T\left(n\right) \qquad (2.61)$$

where $w_P\left(t\right) = \text{rect}\left(\frac{t}{T_I}\right)$ is the transmitted signal support in the fast time domain and $w_n\left(n\right)$ is the slow time window determined by the observation time.

The contribution received by the radar from the kth scatterer located at $\left(x_{1_k}, x_{2_k}\right)$ is a delayed and scaled replica of the transmitted signal,

$$s_R\left(t, n; x_{2_k}\right) = \sigma_k s_T\left(t - \frac{2R_k(n)}{c}, n\right) \qquad (2.62)$$

where $R_k(n)$ denotes the radar scatterer distance. After demodulation, the received signal is given by Equation (2.63)

$$s_R\left(t, n; x_{2_k}\right) = \sigma_k e^{-j\frac{4\pi}{\lambda}R_k(n)} e^{j\pi\mu\left(t - \frac{2R_k(n)}{c}\right)^2} w_P\left(t - \frac{2R_k(n)}{c}\right) w_n\left(n\right)$$
$$(2.63)$$

where $R_k\left(n\right) = \sqrt{\left(R_0 + x_{2_k}\right)^2 + \left(vnT_R - x_{1_k}\right)^2}$ is the distance of the kth scatterer from the radar. The CS algorithm consists of a number of steps which are described in the following paragraphs and represented in Figure 2.16, where for the sake of simplicity, the dependency of the signals from x_{2_k} has been removed.

Step 1: Azimuth Fourier Transform

The first step of the CS algorithm is the RCMC, performed in the fast time/Doppler domain via multiplication of the received signal with a quadratic phase function. Therefore, the first processing step consists of an FT along the slow time domain towards the Doppler frequency domain, which is performed by exploiting the principle of stationary phase [19]. The received signal after azimuth FT can be rewritten as

$$S_R\left(t, \nu; x_{2_k}\right) =$$

$$\sigma_k w_P\left(t - \frac{2R_k\left(\nu\right)}{c}\right) e^{-j\pi\mu_D\left(\nu, x_{2_k}\right)\left(t - \frac{2R_k\left(\nu\right)}{c}\right)^2} e^{-j\frac{4\pi}{\lambda}x_{2_k}B\left(\nu\right)} W_T\left(\nu\right) \quad (2.64)$$

where $W_T(\nu) = FT\left[w_T(n)\right]$ and

$$R_k\left(\nu\right) = \frac{x_{2_k}}{\sqrt{1 - \left(\frac{\lambda\nu}{2v}\right)^2}} = x_{2_k}\left[1 + C_F\left(\nu\right)\right] \quad (2.65)$$

$$C_F\left(\nu\right) = \frac{1}{B\left(\nu\right)} - 1 \quad (2.66)$$

$$B\left(\nu\right) = \sqrt{1 - \left(\frac{\lambda\nu}{2v}\right)^2} \quad (2.67)$$

$$\frac{1}{\mu_D\left(\nu, x_{2k}\right)} = \frac{1}{\mu} + x_{2_k} A\left(\nu\right) \quad (2.68)$$

$$A\left(\nu\right) = \frac{2\lambda\left(B^2\left(\nu\right) - 1\right)}{c^2 B^3\left(\nu\right)} \quad (2.69)$$

As can be easily noticed, the first exponential term in Equation (2.64) denotes a new frequency modulation of the transmitted chirp, which depends on the range of the kth scatterer (x_{2_k}). When this dependency is neglected, then the algorithm leads to defocused images for high squint angles. This frequency modulation is associated with the Doppler chirp rate, $\mu_D\left(\nu, x_{2_k}\right)$, which is related to the chirp rate of the transmitted signal, μ, as shown in Equation (2.68). The term $R_k\left(\nu\right)$ in Equation (2.65) denotes the range migration of the kth scatterer while the *curvature factor*, $C_F\left(\nu\right)$, represents the dependence of the scatterer trajectory from the Doppler frequency. Equation (2.69) shows the *range distortion factor*, which contributes to the secondary range compression (SRC).

Step 2: Chirp Scaling

From Equation (2.65), the range migration in the range/Doppler domain changes according to the range of each target scatterer, x_{2_k}, as represented in Figure 2.17.

In order to perform a range cell migration equalization, i.e., to equalize the range curvature of each range profile in the range/Doppler domain, the received signal is multiplied by a phase signal denoted by

$$H_1\left(t, \nu\right) = e^{-j\pi\mu_D\left(\nu, x_{2_{(REF)}}\right)C_F(\nu)\left(t - \frac{2R_{REF}(\nu)}{2}\right)^2} \quad (2.70)$$

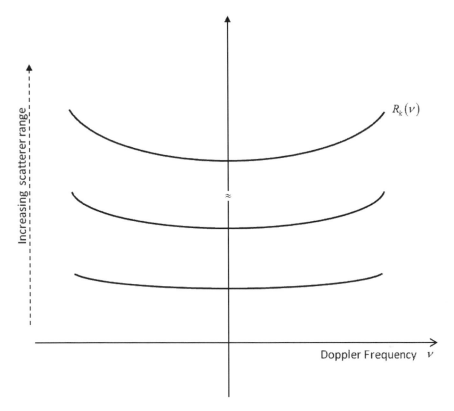

Figure 2.17 Point-like scatterer trajectories representing the behavior of the RCM with respect to the range

where $R_{REF}(\nu) = R_k(\nu)\big|_{x_{2_k}=x_{2_{REF}}} = x_{2_{REF}}\left[1 + C_F(\nu)\right]$ is a reference range used for range cell migration equalization.

The signal after multiplication with the chirp scaling, $H_1(t, \nu)$, is given by

$$S_{R1}(t, \nu; x_{2_k}) = \sigma_k w_P\left(t - \frac{2R_k(\nu)}{c}\right) e^{-j\pi\mu_D\left(\nu, x_{2_{REF}}\right)\left(t - \frac{2}{c}\left(x_{2_k} + C_F(\nu)x_{2_{REF}}\right)\right)^2}$$

$$\cdot\, e^{-j\pi\mu_D\left(\nu, x_{2_{REF}}\right)C_F(\nu)\left(\frac{2}{c}\left(x_{2_k} - x_{2_{REF}}\right)\right)^2} e^{-j\frac{4\pi}{\lambda}x_{2_k}B(\nu)} W_T(\nu)$$

$$(2.71)$$

The equalized range cell migration for the generic k^{th} scatterer, given by $R_{k,EQ}(\nu) = R_{REF}(\nu) - x_{2_{REF}} + x_{2_k}$, is represented in Figure 2.18 by black solid lines.

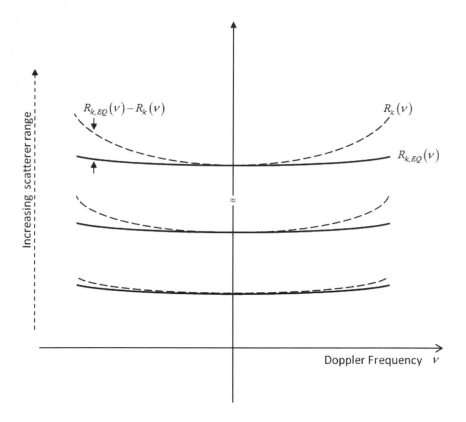

Figure 2.18   Equalized range curvature after chirp scaling operation

Step 3: Range Fourier Transform

At this point, a range FT is performed which highlights the effects of the chirp scaling. The signal after range FT is

$$S_{R2}\left(f,\nu;x_{2_k}\right) = \sigma_k W_P\left(-\frac{f}{\mu'_D\left(\nu,x_{2_{REF}}\right)}\right)$$

$$\cdot e^{\left[-j\frac{4\pi}{\lambda}x_{2_k}B(\nu)-j\pi\mu'_D\left(\nu,x_{2_{REF}}\right)C_F(\nu)\left[\frac{2}{c}\left(x_{2_k}-x_{2_{REF}}\right)\right]^2\right]} \quad (2.72)$$

$$\cdot e^{\left[j\frac{\pi f^2}{\mu'_D\left(\nu,x_{2_{REF}}\right)}\right]}e^{\left[-j\frac{4\pi f}{c}\left[x_{2_k}+C_F(\nu)x_{2_{REF}}\right]\right]}W_T\left(\nu\right)$$

where $\mu'_D\left(\nu,x_{2_{REF}}\right) = \mu_D\left(\nu,x_{2_{REF}}\right)\left[1+C_F\left(\nu\right)\right]$ and $W_P(f) = FT\left[w_P(t)\right]$ where t represents the fast-time variable.

The phase terms in Equation (2.72) can be explained as follows:

> The first one represents an azimuth modulation, constant with respect to the range

The second phase term is a quadratic function of the range frequency, which represents the effective chirp modulation with a modified chirp rate given by $\mu'_D\left(\nu, x_{2_{REF}}\right)$ and is associated to a secondary range migration

The third phase function is linearly dependent on the range frequency and carries information about the correct range position of the scatterer and its range curvature.

Step 4: Range Cell Migration Correction and Range Focus

A range correction term is applied in order to remove the secondary range migration and the range curvature, which is evident in the second and third phase terms of Equation (2.72). The range correction term is given by

$$H_2\left(f, \nu\right) = e^{\left[-j\frac{\pi f^2}{\mu'_D\left(\nu, x_{2_{REF}}\right)}\right]} e^{\left[j\frac{4\pi x_{2_{REF}}}{c}C_F(\nu)f\right]} \tag{2.73}$$

where the first phase term achieves range focus and Secondary Range Compression (SRC) while the second term performs the RCMC. The signal after range correction is given by

$$S_{R3}\left(f, \nu; x_{2_k}\right) = \sigma_k W_P\left(-\frac{f}{\mu'_D\left(\nu, x_{2_{REF}}\right)}\right) e^{\left[-j\frac{4\pi x_{2_k}f}{c}\right]}$$
$$\cdot e^{\left[-j\frac{4\pi}{\lambda}x_{2_k}B(\nu) - j\pi\mu'_D\left(\nu, x_{2_{REF}}\right)C_F(\nu)\left(x_{2_k} - x_{2_{REF}}\right)^2\right]} W_T\left(\nu\right) \tag{2.74}$$

Step 5: Range Inverse Fourier Transform

At this point, a range IFT is performed to focus the signal in the range domain. The signal after range IFT can be rewritten as

$$S_{R4}\left(t, \nu; x_{2_k}\right) \propto \sigma_k \text{sinc}\left[B\left(t - \frac{2x_{2_k}}{c}\right)\right]$$
$$\cdot e^{\left[-j\frac{4\pi}{\lambda}x_{2_k}B(\nu) - j\pi\mu'_D\left(\nu, x_{2_{REF}}\right)C_F(\nu)\left(x_{2_k} - x_{2_{REF}}\right)^2\right]} W_T\left(\nu\right) \tag{2.75}$$

where $B = \frac{\mu T_I}{2\pi}$ is the transmitted signal bandwidth and $\propto$ denotes proportionality since most of the constant terms have been omitted for the sake of simplicity.

As can be easily noted from Equation (2.75), a residual phase term, which results from the chirp scaling operation, is present in the signal. In order to obtain the focused signal, a further phase correction is required, which is performed in the next step.

Step 6: Residual Range Correction and Azimuth Focus

The residual phase term can be removed by multiplying the signal in Equation (2.75) by the following compensation term:

$$H_3\left(\nu\right) = e^{\left[j\frac{4\pi}{\lambda}x_{2_k}B(\nu)+j\pi\mu'_D\left(\nu,x_{2_{REF}}\right)C_F(\nu)\left(\frac{2}{c}\left(x_{2_k}-x_{2_{REF}}\right)\right)^2\right]} \tag{2.76}$$

After residual phase compensation, the signal can be rewritten as

$$S_{R5}\left(t,\nu;x_{2_k}\right) = C_{CS}W_T\left(\nu\right)\operatorname{sinc}\left(B\left(t-\frac{2x_{2_k}}{c}\right)\right) \tag{2.77}$$

where C_{CS} contains all the constant factors resulting from the transformation.

Step 7: Azimuth Inverse Fourier Transform

At this stage, the signal in the fast time/slow time domain can be found by easily applying an azimuth IFT to the signal in Equation (2.77), which leads to

$$s_{R6}\left(t,n;x_{2_k}\right) \propto w_T\left(n\right)\operatorname{sinc}\left(B\left(t-\frac{2x_{2_k}}{c}\right)\right) \tag{2.78}$$

Step 8: Range and Azimuth Scaling

The last step to be performed is a range/azimuth scaling, which can be done by exploiting the following relationships:

$$\begin{cases} t = \frac{2x_2}{c} \\ nT_R = \frac{x_1}{v} \end{cases} \tag{2.79}$$

At this point, the SAR image is given by the modulus of the signal in Equation (2.78)

$$I_{CS}\left(x_1,x_2\right) = \left|s_{R6}\left(t,n;x_{2_k}\right)\right|\Big|_{\substack{t = \frac{2x_2}{c} \\ nT_R = \frac{x_1}{v}}} \tag{2.80}$$

It is worth noting that the image of a given scatterer is always a 2D sinc-like function as in Equation (2.28).

2.4 CONCLUSIONS

This chapter dealt with the SAR imaging topic. SAR is a widely used technique in Earth's observation and remote sensing applications. A number of methods for the image formation process can be found in literature and the most widely used have been recalled in this chapter. An overall view of the SAR system was given first, in which the differences among the SAR acquisition modes were explained. A SAR signal model was introduced and the most widely used image formation techniques described.

REFERENCES

1. Dale A. Ausherman, Adam Kozma, Jack L. Walker, Harrison M. Jones, and Enrico C. Poggio. Developments in Radar Imaging. *Aerospace and Electronic Systems, IEEE Transactions on*, 20(4):363–400, July 1984.
2. J.L. Bauck and W.K. Jenkins. Convolution-backprojection image reconstruction for bistatic Synthetic Aperture Radar. *Circuits and Systems, 1989., IEEE International Symposium on*, 3:1512–1515, May 1989.
3. F. Berizzi, E.D. Mese, M. Diani, and M. Martorella. High-resolution ISAR imaging of maneuvering targets by means of the range instantaneous doppler technique: modeling and performance analysis. *Image Processing, IEEE Transactions on*, 10(12):1880–1890, Dec. 2001.
4. W.G. Carrara, R.S. Goodman, and R.M. Majewski. *Spotlight Synthetic Aperture Radar: Signal Processing Algorithms*. Artech House Remote Sensing Library. Artech House, 1995.
5. I.G. Cumming and F.H. Wong. *Digital Processing of Synthetic Aperture Radar Data: Algorithms and Implementation*. Number v. 1 in Artech House Remote Sensing Library. Artech House, 2005.
6. J.C. Curlander and R.N. McDonough. *Synthetic Aperture Radar: Systems and Signal Processing*. Wiley Series in Remote Sensing and Image Processing. Wiley, 1991.
7. M.D. Desai and W.K. Jenkins. Convolution backprojection image reconstruction for spotlight mode Synthetic Aperture Radar. *Image Processing, IEEE Transactions on*, 1(4):505–517, Oct. 1992.
8. G. Franceschetti and R. Lanari. *Synthetic Aperture Radar Processing*. Electronic engineering systems series. Taylor & Francis, 1999.
9. Floyd M. Henderson, Anthony J. Lewis, and Robert A. Ryerson, editors. *Principles and Applications of Imaging Radar*. Manual of Remote Sensing. Wiley, New York, 1998. Manual of Remote Sensing edited by Robert A. Ryerson.
10. C.V. Jakowatz. *Spotlight-Mode Synthetic Aperture Radar: A Signal Processing Approach*. Springer, 1996.
11. D. Massonnet and J.C. Souyris. *Imaging with Synthetic Aperture Radar*. EFPL Press, 2008.
12. A. Moreira and Yonghong Huang. Airborne SAR processing of highly squinted data using a chirp scaling approach with integrated motion compensation. *Geoscience and Remote Sensing, IEEE Transactions on*, 32(5):1029–1040, Sept. 1994.

13. A. Moreira, J. Mittermayer, and R. Scheiber. Extended chirp scaling algorithm for air- and spaceborne SAR data processing in stripmap and ScanSAR imaging modes. *Geoscience and Remote Sensing, IEEE Transactions on*, 34(5):1123–1136, Sept. 1996.

14. A Moreira, P. Prats-Iraola, M. Younis, G. Krieger, I Hajnsek, and K.P. Papathanassiou. A tutorial on Synthetic Aperture Radar. *Geoscience and Remote Sensing Magazine, IEEE*, 1(1):6–43, March 2013.

15. Jr. Munson, D.C., J.D. O'Brien, and W. Jenkins. A tomographic formulation of spotlight-mode Synthetic Aperture Radar. *Proceedings of the IEEE*, 71(8):917–925, Aug. 1983.

16. Jr. Munson, D.C., J.D. O'Brien, and W. Jenkins. A tomographic formulation of spotlight-mode Synthetic Aperture Radar. *Proceedings of the IEEE*, 71(8):917–925, Aug. 1983.

17. R.K. Raney, H. Runge, R. Bamler, I.G. Cumming, and F.H. Wong. Precision SAR processing using chirp scaling. *Geoscience and Remote Sensing, IEEE Transactions on*, 32(4):786–799, July 1994.

18. M.A. Richards. *Fundamentals of Radar Signal Processing*. McGraw-Hill Education, 2005.

19. M. Soumekh. *Fourier Array Imaging*. PTR Prentice-Hall, 1994.

20. M. Soumekh. *Synthetic Aperture Radar Signal Processing with MATLAB Algorithms*. Wiley-Interscience Publication. Wiley, 1999.

3 ISAR Processing

M. Martorella, F. Berizzi, E. Giusti, and A. Bacci

CONTENTS

This chapter focuses on ISAR signal processing techniques. As already stated in Chapter 1, ISAR is another way to face the problem of forming high resolution radar images of targets. Differently from SAR, ISAR techniques exploit the target motion to form e.m. images of the illuminated targets. Since the targets are typically non-cooperative, their motions are unknown. Then, to coherently process the targets echoes, the relative motion between the radar and the target must be estimated first by means of *autofocusing* techniques. This is the key point that differentiates SAR from ISAR.

The reminder of the chapter is organized as follows. Section 3.1 describes the ISAR signal model. The ISAR image formation chain is instead the objective of Section 3.2. Section 3.3 suggests how to deal with the ISAR parameters setting in a real application. Finally, the main points of this chapter are summarized in Section 3.4.

3.1 ISAR SIGNAL MODEL

In this section the ISAR signal model for a monostatic configuration will be presented.

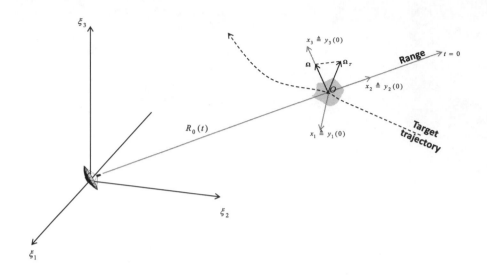

Figure 3.1 ISAR geometry

Consider the typical ISAR geometry shown in Figure 3.1 in which the transmitter and the receiver are colocated in the origin of the Cartesian reference system T_ξ. As stated in Chapter 1, the signal in the frequency/slow time domain at the output of the matched filter can be expressed as

$$S(f,n) = C \cdot \text{rect}\left(\frac{n}{N}\right) |S_T(f,n)|^2 \int_V f'(\mathbf{x}) e^{-j2\pi f \tau(\mathbf{x},n)} \, d\mathbf{x} \qquad (3.1)$$

where in conventional active radar $|S_T(f)|^2 \simeq \text{rect}\left(\frac{f-f_0}{B}\right)$ and $\text{rect}\left(\frac{n}{N}\right) = u\left(n + \frac{N}{2}\right) - u\left(n - \frac{N}{2} - 1\right)$ defines the signal support in the frequency/slow time domain. The function $u(\cdot)$ yields 1 when $(\cdot) > 0$. $\tau(\mathbf{x},n)$ is the delay time of a scatterer placed in $\mathbf{x} = (x_1, x_2, x_3)^T$ with respect to the reference system T_x and is expressed as follows:

$$\tau(\mathbf{x},n) = \frac{2R(\mathbf{x},n)}{c} \qquad (3.2)$$

where $R(\mathbf{x},n)$ is the distance between the radar and the scatterer placed in $\mathbf{x}$ in the nth sweep, c is the speed of light in a vacuum and $\mathbf{\Omega}_T(n)$ and $\mathbf{\Omega}(n)$ are, respectively, the total target rotation vector and the effective target rotation vector, which is obtained as the projection of $\mathbf{\Omega}_T(n)$ onto the plane orthogonal to the radar *line of sight* (LoS).

By assuming that the radar target distance is much greater than the size of the target (*straight iso-range* approximation), the term $R(\mathbf{x},n)$ can be approximated as follows:

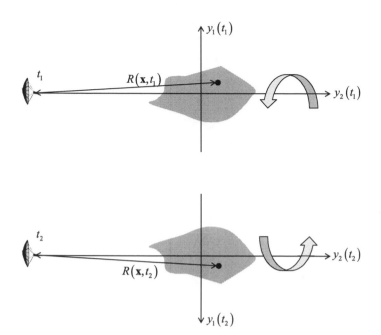

Figure 3.2 Rotation along the LoS

$$R\left(\mathbf{x}, n\right) \simeq R_0\left(n\right) + \mathbf{x}^T \cdot \mathbf{i}_{LoS}\left(n\right) \tag{3.3}$$

where $R_0\left(n\right)$ is the distance between the radar and a reference point on the target at the time instant nT_R, $\mathbf{x}$ is the column vector that identifies a scatterer on the target and $\mathbf{i}_{LoS}$ is the column unit vector that identifies the radar LoS.

It is worth pointing out that Equation (3.3) holds true only when considering the target as a rigid body. In this case, its motion with respect to the radar can be considered as the superimposition of the translational motion of a reference point and a rotation vector applied to this reference point. The rotation vector takes into account both the rotation due to the translation and the proper target rotation. It is worth pointing out that, since the radar can measure only the distance variation along the LoS, any rotation around the y_2 axis does not produce any effect. Then, the only component to be considered is $\mathbf{\Omega}(n)$, as it contributes to the synthetic aperture formation. This result is evident by observing Figure 3.2 in which the case in which the target undergoes a rotation around the LoS is represented. In such a case, the distance between the radar and each scatterer on the target does not change, $R\left(\mathbf{x}, t_1\right) = R\left(\mathbf{x}, t_2\right)$. Then, such a rotation does not produce any effect from the radar point of view.

Let us assume that the effective rotation vector is constant during the overall observation time, i.e., $\mathbf{\Omega}\left(n\right) \approx \mathbf{\Omega}$ for $\left|nT_R\right| < T_{obs}$, and consider the geometry depicted in Figure 3.1 in which the axis x_3 is oriented along $\mathbf{\Omega}$. Under this assumption, the inner product $y_2\left(\mathbf{x}, n\right) = \mathbf{x} \cdot \mathbf{i}_{LoS}\left(n\right)$ can be expressed in a closed form by solving the differential equation system expressed as follows:

$$\dot{\mathbf{y}}(n) = \mathbf{\Omega}_T \times \mathbf{y}(n) \tag{3.4}$$

with the initial condition $\mathbf{y}(0) = \mathbf{x}$.

The resulting closed form solution is

$$\mathbf{y}(n) = \mathbf{a} + \mathbf{b}\cos\left(\Omega_T n T_R\right) + \frac{\mathbf{c}}{\Omega_T}\sin\left(\Omega_T n T_R\right) \tag{3.5}$$

where

$$\mathbf{a} = \frac{\mathbf{\Omega}_T \cdot \mathbf{x}}{\Omega_T^2}\mathbf{\Omega}_T$$
$$\mathbf{b} = \mathbf{x} - \frac{\mathbf{\Omega}_T \cdot \mathbf{x}}{\Omega_T^2}\mathbf{\Omega}_T$$
$$\mathbf{c} = \mathbf{\Omega}_T \times \mathbf{x} \text{ and } \Omega_T = |\mathbf{\Omega}_T|$$
$$\Omega_T = |\mathbf{\Omega}_T|$$

with the above-mentioned choice of the reference system, $\mathbf{\Omega}_T = \left(0, \Omega_{T_2}, \Omega\right)$ (with respect to T_y) where the component Ω_{T_2} does not produce any effect as stated above.

Equation (3.5) can be reasonably approximated by its first order Taylor series around $t = 0$ as follows:

$$\mathbf{y}(n) \approx \mathbf{a} + \mathbf{b} + \mathbf{c}\left(n T_R\right) = \mathbf{x} + \mathbf{c}\left(n T_R\right) \tag{3.6}$$

resulting in:

$$y_2\left(\mathbf{x}, n\right) = x_2 + \Omega n T_R x_1 \tag{3.7}$$

where $\Omega = |\mathbf{\Omega}|$.

By substituting Equation(3.6) in Equation (3.1) and by defining $W(f, n) = C \cdot \text{rect}\left(\frac{f - f_0}{B}\right)\text{rect}(n/N)$ (as defined in Equation (1.17)), the received signal can be rewritten as

$$S(f, n) = W(f, n)\int_V f'(\mathbf{x})e^{-j4\pi\frac{f}{c}\left(R_0(n) + x_2 + \Omega x_1 n T_R\right)}d\mathbf{x} \tag{3.8}$$

The same result can be obtained in a different way considering that the LoS unit vector in the nth sweep in the T_x reference system is given by

$$\mathbf{i}_{LoS}(n) = \begin{bmatrix} \sin\left(\Omega n T_R\right) \\ \cos\left(\Omega n T_R\right) \\ 0 \end{bmatrix} \tag{3.9}$$

As a consequence, the inner product in Equation (3.3) results in

$$y_2\left(\mathbf{x}^T, n\right) = \mathbf{x} \cdot \mathbf{i}_{LoS}(n) = x_2\cos\left(\Omega n T_R\right) + x_1\sin\left(\Omega n T_R\right) \tag{3.10}$$

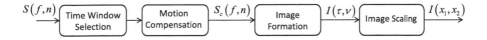

Figure 3.3 ISAR processing chain

Equation (3.1) can therefore be written as

$$S(f,n) = W(f,n) \iint\limits_{x_1,x_2} f(x_1,x_2) \, e^{-j\frac{4\pi f}{c}[R_0(n)+x_2\cos(\Omega n T_R)+x_1\sin(\Omega n T_R)]} dx_1 dx_2$$

(3.11)

where $f(x_1,x_2) = \int_{x_3} f'(\mathbf{x}) \, dx_3$ is the projection of the target's reflectivity function onto the image projection plane (IPP).

By assuming the total aspect angle variation is small during the observation time, i.e., $\Delta\vartheta = \Omega T_{obs} \ll 1$, Equation (3.11) becomes Equation (3.8).

3.2   ISAR IMAGE FORMATION CHAIN

The steps to be performed to obtain an ISAR image are summarized in the functional block diagram shown in Figure 3.3. A time interval of the overall observation within which the effective rotation vector can be assumed constant time is selected first by means of the *time windowing selection*. This operation is important because when the radar observes the target for long time intervals, this assumption does not hold true and the model in Equation (3.8) cannot be applied. Through the time windowing process, a subset of data for which the effective rotation vector is constant is properly selected to form the ISAR image. After that, the motion compensation operation is performed. This operation aims at removing the phase term $R_0(n)$ that accounts for the target radial motion. After motion compensation, the image formation step is performed via a range-Doppler (RD) approach. This operation provides the ISAR image in the delay time/Doppler coordinates (τ, ν). Finally, in order to better asses the target size, a scaling operation to spatial coordinates is needed.

All these processing steps will be extensively described in the following sections.

3.2.1   IMAGE FORMATION

The ISAR image is formed via the RD approach. Consider the received signal expressed in Equation (3.11). Assuming that the phase term due to the translational motion $R_0(n)$ is perfectly compensated during the ISAR image

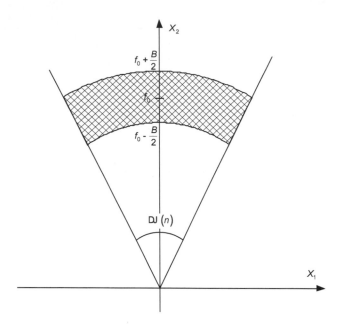

Figure 3.4 Spatial frequency domain

formation processing, the motion compensated signal is expressed as follows:

$$S_C\left(X_1, X_2\right) = W(X_1, X_2) \iint\limits_{x_1,x_2} f\left(x_1,x_2\right) e^{-j2\pi(x_1 X_1 + x_2 X_2)} dx_1 dx_2 \qquad (3.12)$$

where (X_1, X_2) are the spatial frequencies defined as

$$\begin{cases} X_1(f,n) = \frac{2f}{c} \sin\left(\Omega n T_R\right) \\ X_2(f,n) = \frac{2f}{c} \cos\left(\Omega n T_R\right) \end{cases} \qquad (3.13)$$

From Equation (3.13) it is obvious that the motion compensated signal corresponds to the windowed Fourier transform (FT) of the projected target's reflectivity function. The term $W(X_1, X_2)$ defines the region (a sector of annulus) in the spatial frequencies domain in which the signal is defined, see Figure 3.4.

As stated above, by assuming that the total aspect angle variation is small, i.e., $\Delta\vartheta = \Omega T_{obs} \ll 1$, the spatial frequencies can be approximated as follows:

$$\begin{cases} X_1(n) = \frac{2f_0}{c}\left(\Omega n T_R\right) \\ X_2(f) = \frac{2f}{c} \end{cases} \qquad (3.14)$$

It is worth noting that for X_1 the frequency f has been substituted by the central frequency f_0, being the result of the approximation of the polar

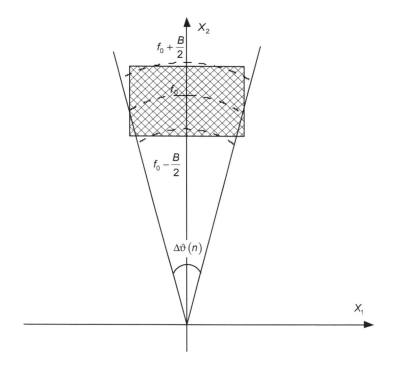

Figure 3.5 Spatial frequency domain with small aspect angle variation

domain in Figure 3.4 with a rectangular window that intersects the angular sector at the coordinates $X_2 = \frac{2f_0}{c}$. It should be noted that this approximation is the one that leads to the minimum error as it can be inferred by examining Figure 3.5 where both the polar and the rectangular domains are represented.

By assuming that the target is composed by K point-like scatterers and neglecting all the interactions among the scatterers, the target reflectivity function can be expressed as follows:

$$f(x_1, x_2) = \sum_{k=1}^{K} \sigma_k \delta(x_1 - x_{1_k}, x_2 - x_{2_k}) \tag{3.15}$$

where x_{1_k}, x_{2_k} and σ_k are the cross-range coordinate, the range coordinate and the reflectivity value of the $k^{(th)}$ scatterer, respectively.

By substituting Equation (3.15) in Equation (3.12) the received signal after motion compensation results

$$S_C(X_1, X_2) = W(X_1, X_2) \sum_{k=1}^{K} \sigma_k e^{-j2\pi(X_1 x_{1_k} + X_2 x_{2_k})} \tag{3.16}$$

The ISAR image can be obtained by applying a 2D-FT to Equation (3.16) and can be expressed as

$$I(x_1, x_2) = w(x_1, x_2) \otimes \otimes f(x_1, x_2) = \sum_{k=1}^{K} \sigma_k w\left(x_1 - x_{1_k}, x_2 - x_{2_k}\right) \quad (3.17)$$

By substituting Equation (3.14) in Equation (3.16), the received signal after motion compensation becomes:

$$S_C(f, n) = W(f, n) \sum_{k=1}^{K} \sigma_k e^{-j2\pi(f\tau_k + nT_R \nu_k)} \quad (3.18)$$

where the delay-time and the Doppler coordinates are defined as:

$$\begin{aligned} \tau &= \frac{2x_2}{c} \\ \nu &= \frac{2f_0 \Omega x_1}{c} \end{aligned} \quad (3.19)$$

The ISAR image in the delay time/Doppler domain, $I(\tau, \nu)$ can then be obtained via a two-dimensional Fourier transform of Equation (3.18) as follows:

$$I(\tau, \nu) = w(\tau, \nu) \otimes \otimes \sum_{k=1}^{K} \sigma_k \delta\left(\tau - \tau_k, \nu - \nu_k\right) = \sum_{k=1}^{K} \sigma_k w\left(\tau - \tau_k, \nu - \nu_k\right)$$
$$(3.20)$$

where

$$w(\tau, \nu) = \text{2D-FT}\left[W(f, t)\right] = BT_{obs} \text{sinc}\left(\tau B\right) \text{sinc}\left(\nu T_{obs}\right) e^{-j2\pi f_0 \tau} \quad (3.21)$$

and 2D-FT$[\cdot]$ is the two-dimensional Fourier transform operator, $\text{sinc}(x) = \frac{\sin(\pi x)}{\pi x}$ and $\nu \in \left[-\frac{PRF}{2}, \frac{PRF}{2}\right]$. $w(\tau, \nu)$ defines the system point spread function (PSF) and can be used to find the resolution in the delay time/Doppler dimensions, i.e., δ_τ and δ_ν, defined as the first null of the two *sinc* functions in both directions

$$\begin{aligned} \delta_\tau &= \frac{1}{B} \\ \delta_\nu &= \frac{1}{T_{obs}} \end{aligned} \quad (3.22)$$

It is worth pointing out that Equation (3.20) represents the ISAR image in the delay time/Doppler domain. To obtain the target ISAR image in the spatial coordinates (range/cross-range) a scaling operation is needed. This scaling operation can be performed by exploiting Equation (3.19) as follows:

$$\begin{cases} x_1 = \frac{c}{2f_0 \Omega} \nu \\ x_2 = \frac{c}{2} \tau \end{cases} \quad (3.23)$$

and, as a consequence, the spatial resolutions can be expressed as:

$$\begin{cases} \delta_{x_1} = \frac{c}{2f_0 \Omega T_{obs}} \\ \delta_{x_2} = \frac{c}{2B} \end{cases} \tag{3.24}$$

Equation (3.23) shows that the scaling operation from the delay time to the range coordinate is straightforward while this is not true for the scaling operation from the Doppler frequency to the cross-range coordinates. In fact, this coordinates conversion involves the knowledge of the modulus of the effective rotation vector, Ω, which depends on the motion of the non-cooperative moving target and, as a consequence, cannot be assumed known in the ISAR scenario.

As can be noted from Equation (3.24), the resolution becomes finer as the aspect angle variation increases. This is in contrast with the small aspect angle variation assumption which is mandatory for the applicability of the range Doppler image formation algorithm. This hypothesis determines an upper bound in the cross-range resolution that can be achieved with this algorithm.

Another important consideration concerns the definition of the IPP where the target is represented. As stated above, the reference system T_x has been chosen so that the x_2 axis is oriented along the LoS for $t = 0$ and the x_3 axis is parallel to the vector Ω that produces the aspect angle variation. Since any rotation around the LoS does not produce any variation in the aspect angle, it is quite obvious that Ω is the projection onto the plane orthogonal to the LoS of the total rotation vector Ω_T and can be expressed as follows:

$$\Omega(n) = \mathbf{i}_{LoS} \times [\Omega_T(n) \times \mathbf{i}_{LoS}] \tag{3.25}$$

Therefore, the cross-range axis $\mathbf{x}_1$ is defined as:

$$\mathbf{x}_1 = \Omega \times \mathbf{i}_{LoS} \tag{3.26}$$

where the dependency on n is dropped out because the IPP can be defined only if Ω is constant during the observation time. It is evident that, as it happens for the cross-range scaling operation, the IPP depends on the unknown quantity Ω so that the IPP cannot be a priori predicted leading to some difficulties in the ISAR image interpretation.

It is worth pointing out that the above-mentioned RD technique is based on the assumption that the Doppler frequency of each scatterer, relative to the focusing point, is constant during the CPI (coherent processing interval). This hypothesis usually holds true for low spatial resolution (of the order of meter) and in the case in which the target does not undergo fast maneuvers and/or is affected by significant oscillation such as roll, pitch and yaw. In the case in which very high resolutions are required, longer CPIs are needed and the Doppler frequency becomes time-varying. This can happen also when the target undergoes angular motions, as in the case of ships. In this case the first order Taylor approximation in Equation (3.6) and (3.7) is not valid. To solve this problem, the *time frequency transforms* (TFTs) is used instead of the RD approach to form ISAR images. TFTs are, in fact, suitable for the analysis of non-stationary signals. An example of a TFT- based technique is the range instantaneous Doppler (RID) approach described in [2].

3.2.2   TIME WINDOW SELECTION

As stated before, the applicability of the RD technique relies on the assumption that the effective rotation vector is constant during the observation time, i.e., $\Omega(n) \approx \Omega$. However, the target's own motion may induce a non-uniform target rotation vector. In particular, maritime targets can undergo very complex motions (roll, pitch and yaw) induced by the sea surface. In order to minimize the target's rotation variations, the CPI can be controlled via a time-windowing approach.

Typically, a fixed window length can be set and a sliding window processing along the whole observation time is applied to obtain a sequence of ISAR images of the targets. Among the obtained frames, the more suitable for target classification and recognition is selected.

It is worth pointing out that, in order to obtain good classification and recognition performances, the most important requirement is a good level of focus, as the target details would then appear sharper than in defocused images. As stated above, the image cross-range resolution is inversely proportional to the total aspect angle variation and, as a consequence, to the considered CPI so that longer CPI involves a finer resolution in the cross-range dimension. On the other hand, longer CPIs increase the chance that the target rotation vector may not be considered constant and therefore producing a defocused image. It is quite obvious that a trade-off must be identified to obtain a well focused and a high resolved image at the same time.

An automatic time windowing technique is proposed in [9] and will be here recalled for the sake of clarity. Specifically, the position of the window along the whole observation time and its length are automatically selected in order to obtain the image with the highest focus. The concept can be explained referring to Figure 3.6, in which the data is refereed to be distributed along the time axis n (n denotes the sweep index at time $t = nT_R$). The considered temporal window is defined by two parameters, namely n_w that identifies the window's position and N_w that denotes the window's length in number of sweeps, so that the temporal length of the windows is $T_w = N_w T_R$.

The measure of the image focus is made through the image contrast (IC) which is assumed to be maximum when the window position (n_w) and the window length (N_w) identify a subset of data where the conditions of constant rotation vector and resolution are optimal in terms of image focus.

In particular, the optimal time window is obtained by maximizing the IC with respect to the pair (n_w, N_w). This problem can be mathematically formulated as follows:

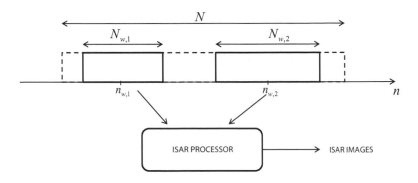

Figure 3.6 Time windowing concept

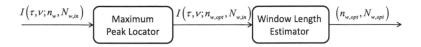

Figure 3.7 Optimal time window estimator

$$(n_{w,opt}, N_{w,opt}) = \underset{(n_w, N_w)}{\arg\max} \left(IC\left(n_w, N_w\right) \right) \qquad (3.27)$$

It should be noted that the variables (n_w, N_w) are both discrete so that the problem in Equation (3.27) is a discrete optimization problem. Specifically, such a problem can be classified as a non-linear Knapsack Problem [11]. The solution proposed in [9] is based on a double linear search, which can be briefly described by the following steps:

1. Contrast maximization with respect to n_w for a given guess $N_{w,in}$. Let $n_{w,opt}$ be the solution of such maximization;

2. Optimization with respect to N_w with $n_w = n_{w,opt}$.

The procedure is depicted in Figure 3.7 for the sake of clarity. The justification of this procedure is heuristic. It can be observed both in simulated and real ISAR data of several types of targets that the position of the optimal time-window is almost independent on the window length and, as a consequence, the IC peak position along the slow-time axis, n, does not change by changing N_w. This can be physically explained by considering that the target own motion is characterized by regular motions at given times and less regular motion at other times. For example, a ship usually undergoes regular pitch and roll motions due to the sea surface state. This regular motion may be disturbed by an incoming wave that generates complex motions which cause the rotation vector to change rapidly. For this reason, the estimation of $n_{w,opt}$ and $N_{w,opt}$ can be performed separately.

3.2.3 MOTION COMPENSATION

As stated in Section 3.2.1, the radial motion compensation is a fundamental step in forming well-focused ISAR images. This step is typically termed *image autofocus* and consists of removing the phase term $-\frac{4\pi f}{c} R_0(n)$ in Equation (3.8). This term depends on the radial motion of the reference point. When no external data are available, this operation must be performed only by using the radar received signal and, for this specific reason, the image focusing process is called ISAR autofocus.

Several techniques have been developed by the ISAR research community, each of them showing pros and cons. Such techniques can be classified as parametric and non-parametric methods [1]. Parametric methods need a parametric model of the radar received signal while non-parametric techniques do not make any assumption. Some autofocus techniques are detailed in the next subsections.

3.2.3.1 Image Contrast Based Autofocus

The *image contrast based autofocus* (ICBA) is a parametric technique that is based on the obvious concept that a higher value of IC corresponds to a more focused image [10]. This autofocus technique aims at removing the term $R_0(n)$ via an iterative estimation of the motion parameters. For a relatively small observation time, $T_{obs} = NT_R$, and relatively smooth target motions, the radar-target residual distance can be expressed by exploiting a Lth polynomial model as follows:

$$R_0(n) = \sum_{l=0}^{L} \frac{1}{l!} \alpha_l (nT_R)^l \qquad (3.28)$$

in which α_l are the above-mentioned *focusing parameters* which can be grouped in a vectorial form as $\boldsymbol{\alpha} = [\alpha_0, \alpha_1, \cdots, \alpha_{L-1}]^T$.

The estimation of $R_0(n)$ results in the estimation of the target motion parameters via the maximization of the IC with respect to $\boldsymbol{\alpha}$ as:

$$\hat{\boldsymbol{\alpha}} = \arg\max_{\boldsymbol{\alpha}} [IC(\boldsymbol{\alpha})] \qquad (3.29)$$

where the IC is defined as follows:

$$IC(\boldsymbol{\alpha}) = \frac{\sqrt{A_v \left[[I(\tau, \nu, \boldsymbol{\alpha}) - A_v [I(\tau, \nu, \boldsymbol{\alpha})]]^2 \right]}}{A_v [I(\tau, \nu, \boldsymbol{\alpha})]} \qquad (3.30)$$

and where $A_v [\cdot]$ denotes the average operation over the variables τ and ν, $I(\tau, \nu, \boldsymbol{\alpha})$ is the ISAR image magnitude or intensity (power) after motion compensating the target's translational motion by using α as focusing parameters and is expressed as follows:

$$I\left(\tau, \nu, \alpha\right) = \left|\text{2D-FT}\left[S\left(f, n\right) e^{j\frac{4\pi f}{c}R_0\left(\alpha, n\right)}\right]\right|^p \tag{3.31}$$

Referring to Equation (3.31), $p = 1$ in the case of image amplitude and $p = 2$ in the case of image intensity. $R_0\left(\alpha, n\right)$ is the radial motion model evaluated with the α motion parameters.

3.2.3.2 Image Entropy Based Autofocus

A similar approach to the one proposed in Section 3.2.3.1 can be devised by substituting the IC with the *image entropy* (IE) expressed as follows:

$$IE\left(\alpha\right) = -\iint \ln\left(\bar{I}\left(\tau, \nu, \alpha\right)\right) \bar{I}\left(\tau, \nu, \alpha\right) d\tau d\nu \tag{3.32}$$

where $\bar{I}\left(\tau, \nu, \alpha\right) = \frac{I(\tau, \nu, \alpha)}{A_v[I(\tau, \nu, \alpha)]}$.

As for the IC, the IE is a good indicator of the image focus. Unlike the IC, the smaller IE value the better the image focus.

3.2.3.3 Dominant Scatterer Autofocus

The *dominant scatterer autofocus* (DSA), also known as the *hot spot* (HS), is a non-parametric two stages technique. The first stage consists of a rough alignment before applying the phase compensation as a second step. This technique is derived from the studies in two different areas of research, namely the time delay estimation [3] and the adaptive beamforming [12]. A complete description of this algorithm can be found in [4] while a brief overview is reported here.

After measuring and storing the complex envelopes of the echo samples, high resolution range profiles in the nth sweep, $S\left(\tau, n\right)$, can be obtained via Fourier transforming Equation (3.1) along the frequency dimension. A cross-correlation shift operation can be performed between successive range profiles in order to obtain a rough range profile alignment. Let $S_a\left(\tau, n\right)$ be the roughly aligned profiles expressed as follows:

$$S_a\left(\tau, n\right) = A\left(\tau, n\right) e^{j\phi\left(\tau, n\right)} \tag{3.33}$$

where τ denotes a range cell (delay time) and n is the pulse number and $A\left(\tau, n\right) \propto \left|S_a\left(\tau, n\right)\right| e^{j\phi_A}$, where $e^{j\phi_A\left(\tau\right)}$ is a phase term independent on the slow-time variable, n.

Then, a search along the range coordinate is performed in order to find a dominant and stable scatterer. The range cell where such scatterer is located is the reference range denoted by τ_0 and the corresponding range profile is given by

$$S_a\left(\tau_0, n\right) = S_0\left(n\right) = A\left(\tau_0, n\right) e^{j\phi_0(n)} \tag{3.34}$$

The τ_0 value is found by measuring the normalized echo variance in each range cell and is determined to be the range for which the variance value is minimum. This approach relies on the assumption that a dominant scatterer with a large radar cross-section exists and, therefore, the measured phase can be attributed to the phase generated by one point scatterer. The next step performs a phase compensation by using the phase history of $S_0(n)$. For the range cell data corresponding to τ_0 the result is expressed as follows:

$$S_{0,c}(n) = A\left(\tau_0, n\right) \approx A \tag{3.35}$$

Applying the same operation to the other range cells, motion compensation is achieved and the result is expressed as

$$S_C\left(\tau, n\right) = A(\tau, n) e^{j[\phi(\tau, n) - \phi_0(n)]} \tag{3.36}$$

This algorithm is also known as *minimum variance algorithm* (MVA) due to the criterion used to choose the reference range cell.

A more robust version of this algorithm, which combines the echoes from several range cells has been proposed in [4] and it is named *multiple scatterer algorithm* (MSA). It is based on an average operation performed on the phase differences of M_c reference scatterer (after phase unwrapping) in order to provide the phase correction. Typically, three reference range bins ($M_c = 3$) are sufficient to produce a focused image. Mathematical details are given below. Let the mth reference cell be expressed as:

$$S_m\left(n\right) = A\left(\tau_m, n\right) e^{j\phi_m(n)} \tag{3.37}$$

and let M_c be the number of selected range cells. The estimation of the phase history to be compensated, $\phi_0(n)$, is

$$\phi_0(n) = \frac{1}{M_c} \sum_{m=1}^{M_c} \phi_m(n) \tag{3.38}$$

Phase compensation is then performed as expressed in Equation (3.35) and (3.36). The DSA algorithm is summarized in the functional block depicted in Figure 3.8.

3.2.3.4 Phase Gradient Autofocus

The *Phase Gradient Autofocus* (PGA) can be seen as an extension of the DSA algorithm proposed in [4]. It also exploits the information that remains in those range cells discarded by the *minimum variance* (MV) range selection. The PGA algorithm replaces the phase compensation approach used in DSA with a solution based on the *maximum likelihood* (ML) estimator. The ML

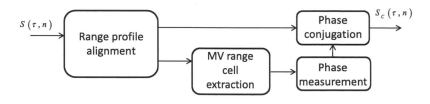

Figure 3.8 DSA algorithm functional block diagram

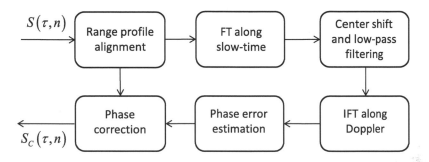

Figure 3.9 PGA functional block diagram

estimator theoretically uses the information contained in all the range cells. However, only the range cells in which the SNR is high enough are considered to improve the PGA performance.

The functional block diagram that summarizes the PGA algorithm is depicted in Figure 3.9.

A range Doppler image is formed and a rough range alignment is performed. Then, the peak value in each range cell, which is supposed to be the return from the dominant scatterer, is found along the Doppler dimension and center-shifted and filtered in the Doppler domain (low-pass filter). Each range cell is then transformed back in the pulse domain via an inverse Fourier transform (IFT) to obtain the phase shifted and filtered time histories.

This operation corresponds to isolating the strongest scatterer contribution for each range cell and forcing it to zero-Doppler, which can be interpreted as a way to remove the scatterer radial motion.

Then the phase gradient is estimated by using the ML approach as described as follows.

Let the arbitrary m^{th} range cell time history be represented as follows, where two consecutive samples are considered:

$$
\begin{aligned}
S_a\left(\tau_m, n-1\right) &= A(\tau_m)e^{j\phi(n-1)} + S_n(\tau_m, n-1) \\
S_a\left(\tau_m, n\right) &= A(\tau_m)e^{j\phi(n)} + S_n(\tau_m, n)
\end{aligned}
\tag{3.39}
$$

where τ_m denotes the mth range cell, $A(\tau_m) = |S_a(\tau_m, n-1)| \simeq |S_a(\tau_m, n)|$ denotes the amplitude, which is assumed to be constant for two consecutive samples, $\phi(n)$ is the phase at time n and $S_n(\tau, n)$ is an additive white Gaussian noise sample.

The phase gradient between two consecutive samples can be defined as

$$\Delta\phi = \phi(n) - \phi(n-1) \tag{3.40}$$

The derivation of the ML estimator of the phase gradient is given in [5]. The solution is expressed as follows

$$\Delta\hat{\phi}(n) = \angle \left(\sum_{m=1}^{M_c} S_a(\tau_m, n) S_a^*(\tau_m, n-1) \right) \tag{3.41}$$

where M_c is the number of range cells used in the estimation process and where the symbol $\angle(\cdot)$ indicates the phase of a complex number.

The phase correction term is then obtained by integrating the estimated phase gradient, as follows:

$$\begin{aligned} \hat{\phi}(n) &= \sum_{l=1}^{n} \Delta\hat{\phi}(l) \\ \hat{\phi}(0) &= 0 \end{aligned} \tag{3.42}$$

Finally, the phase correction term is used to compensate the phase error and obtain the compensated signal.

3.2.4 IMAGE SCALING

The ISAR processing generates a two-dimensional high-resolution image of a target in the delay-time/Doppler domain. In order to determine the size of the target it is, however, required to have a fully scaled image in the range/cross-range domain. The scaling along the range dimension can be performed by using the well-known relationship $x_2 = \frac{c\tau}{2}$, where x_2 is the slant-range coordinate and τ is the delay-time. On the other hand, cross-range scaling requires the estimation of the modulus of the target's effective rotation vector, Ω, since the cross-range scaling is performed as $x_1 = \frac{c}{2f_0\Omega}\nu$. In this section an algorithm to obtain a fully scaled ISAR image is described.

Under the assumption that the target rotation vector is constant within the CPI, the chirp rate (which corresponds to the variation rate of the Doppler frequency of a scatterer) produced by the scattering centers is related to the modulus of the target rotation vector. Therefore, each scattering center carries information about the modulus of the target rotation vector through its chirp rate. To demonstrate this we will consider the radial motion compensated signal expressed in Equation (3.11).

The phase term in the integral can be approximated with a second order Taylor polynomial expansion, as follows:

$$\phi\left(f, n; x_1, x_2\right) = -j \frac{4\pi f}{c} \left(x_2 + x_1 \Omega n T_R - \frac{1}{2} \cdot x_2 \Omega^2 (n T_R)^2\right) \tag{3.43}$$

so that the echo of an ideal scatterer located at (x_{1_k}, x_{2_k}), with finite reflectivity function $f\left(x_1, x_2\right) = \sigma_k \delta\left(x_1 - x_{1_k}, x_2 - x_{2_k}\right)$, can be written as follows:

$$S_c(f, n; x_{1_k}, x_{2_k}) = W\left(f, n\right) \sigma_k e^{\left(-j \frac{4\pi f}{c}\left(x_{2_k} + x_{1_k} \Omega n T_R - \frac{1}{2} \cdot x_{2_k} \Omega^2 (n T_R)^2\right)\right)} \tag{3.44}$$

After applying an FT to the signal in Equation (3.44) along the frequency domain, the range compressed signal is obtained, as shown in Equation (3.45):

$$S_c(\tau, n; x_{1_k}, x_{2_k}) = B\sigma_k \mathrm{sinc}\left(B\left(\tau - \frac{2}{c} x_{2_k}\right)\right) \mathrm{rect}\left(\frac{n}{N}\right) \\ e^{-j \frac{4\pi f_0}{c}\left(x_{2_k} + x_{1_k} \Omega n T_R - \frac{1}{2} x_{2_k} \Omega^2 (n T_R)^2\right)} \tag{3.45}$$

It should be noted that the phase term in Equation (3.45) represents the phase of a chirp signal, where $\mu_k = \frac{2 f_0}{c} x_{2_k} \Omega^2$ is usually referred to as *chirp rate*. If a method for perfectly estimating the chirp rate of a given scatterer was available, the following equation could be exploited to estimate the modulus of the effective target rotation vector:

$$\mu = \frac{2 f_0}{c} x_2 \Omega^2 \tag{3.46}$$

In fact, as the scatterer range coordinate x_2 can be readily obtained by measuring the delay-time τ, the effective rotation vector can be obtained by inverting Equation (3.46), as follows:

$$\Omega = \sqrt{\frac{c}{2 f_0 x_2} \mu} \tag{3.47}$$

In practice, a scatterer chirp rate, as well as its range, must be estimated from the received data. Therefore, the estimation of the effective rotation vector magnitude would, in general, be affected by an error. Different techniques for estimating target scattering center chirp rates making use of atomic decomposition [15], CLEAN technique [13, 8] or based on the IC method have been proposed in [7].

Specifically, the method proposed in [7] is considered here, as it has been proven to be the most effective. According to such a method, the signal components of the scattering centers are processed by means of a polynomial Fourier transform (PFT) in order to estimate the chirp rate. The use of such a method, namely local polynomial Fourier transform (LPFT), requires the solution of an optimization problem [14]. It has been largely proven in literature that the image contrast (IC) and the image entropy (IE) are good indexes to assess the quality of an image. Therefore, the proposed algorithm estimates the scatterers chirp rate that maximize the IC.

To make the estimation more accurate and robust, the chirp rates of a number of target scatterers can be measured together with their ranges. The problem of estimating the effective rotation vector magnitude is then transformed into a problem of estimating the slope of a straight line that fits the scatterplot generated by the set of range and chirp rate estimates. One way of solving this problem is to apply a least square error (LSE) approach [6]. The mathematical problem can be set as follows:

$$\hat{\mu}_k = ax_{2_k} + b + \epsilon_k \tag{3.48}$$

where $a = \frac{2f_0}{c}\Omega^2$, b is the μ-intercept of the line and $\hat{\mu}_k$, x_{2_k} and ϵ_k are the chirp rate estimate, the range estimate and the estimation error for the kth scatterer, respectively. Both a and b are unknown values. The LSE problem and its solution for the estimation of a are given in Equation (3.49) and Equation (3.50).

$$(\hat{a}, \hat{b}) = \arg\min_{a,b} \sum_{k=1}^{K} \epsilon_k^2 \tag{3.49}$$

$$\hat{a} = \frac{K \sum_{k=1}^{K} \hat{\mu}_k x_{2_k} - \sum_{k=1}^{K} \hat{\mu}_k \sum_{k=1}^{K} x_{2_k}}{K \sum_{k=1}^{K} (x_{2_k})^2 - \left(\sum_{k=1}^{K} x_{2_k}\right)^2} \tag{3.50}$$

For the sake of clarity, the algorithm steps can be summarized as follows:

1. Image segmentation
2. Chirp rate estimation via LPFT
3. Effective rotation vector estimation

In detail, after motion compensation, an ISAR image obtained by means of the RD technique can be modeled as the composition of several sinc-like functions centered at each scatterer location in the IPP, as shown in Equation (3.51).

$$I(\tau, \nu) = C \sum_{k=1}^{K} \sigma_k \text{sinc}\left[B\left(\tau - \frac{2x_{2_k}}{c}\right)\right] \text{sinc}\left[T_{obs}\left(\nu - \frac{2f_0\Omega x_{1_k}}{c}\right)\right] \tag{3.51}$$

where K is the number of the scattering centers. After taking the absolute value of the ISAR image, a threshold is applied in order to extract the scattering centers. A number of connected groups of pixels are then found by means of a clustering procedure (segmentation task). In this way the backscattered signal coming from a specific scatterer can be selected. Its expression can be written as follows:

$$S_c(\tau_k, n) = \sigma_k e^{-j\frac{4\pi f_0}{c}x_{2_k}} e^{\left[-j\frac{4\pi f_0}{c}\left(x_{1_k}\Omega nT_R - \frac{x_{2_k}}{2}\Omega^2(nT_R)^2\right)\right]} \tag{3.52}$$

By applying a second order LPFT [14] a complex three variables ISAR image can be obtained in the delay time/Doppler/chirp rate domain as follows:

$$S_c(\tau_k, \nu, \mu) = \sigma_k e^{-j\frac{4\pi f_0}{c}x_{2k}} \sum_n e^{\left[-j\pi\left(\left(\nu + \frac{2f_0}{c}x_{1k}\Omega\right)nT_R + \frac{1}{2}\left(\mu - \frac{2f_0}{c}x_{2k}\Omega^2\right)(nT_R)^2\right)\right]}$$

(3.53)

From Equation (3.53) it is possible to estimate the Doppler frequency, ν_k, and the chirp rate, μ_k, by maximizing the IC of $S_c(\tau_k, \nu, \mu)$ with respect to ν and μ.

From the estimated value of μ_k and the scatterer location τ_k the modulus of the effective rotation vector can be estimated via the LSE approach described in Equation (3.50).

3.3   ISAR PARAMETERS SETTING

In a realistic scenario, the signal to work with is fully digital so all the variables introduced in the signal modeling are discrete and can be expressed as:

$$\begin{aligned} f &= f_0 + m_f \Delta_f & m_f &= 0, ..., M_f - 1 \\ \tau &= m\Delta_\tau & m &= 0, ..., M - 1 \\ \nu &= n_\nu \Delta_\nu & n_\nu &= 0, ..., N_\nu - 1 \end{aligned}$$

(3.54)

Since the relationship between (f, n) and (τ, ν) is given by a Fourier transformation, the sampling intervals in the image domain depend on the ISAR signal spectral occupancy, as follows:

$$\begin{aligned} \delta_\tau &= \frac{1}{M\Delta_f} = \frac{1}{B} \\ \delta_\nu &= \frac{1}{NT_R} = \frac{1}{T_{obs}} \end{aligned}$$

(3.55)

where $M = M_f$ and $N = N_\nu$. As a consequence, the sampling interval along the range and cross-range axes can be derived as in Equation (3.56)

$$\begin{aligned} \delta_r &= \frac{c}{2B} \\ \delta_{cr} &= \frac{c}{2f_0 \Omega T_{obs}} \end{aligned}$$

(3.56)

Moreover, the above-mentioned sampling operation in the ISAR signal domain leads to an ambiguity in the ISAR image domain that must be taken into account in the radar design. Specifically, the non-ambiguity region in the cross-range and range dimension, i.e., Δ_{x_1} and Δ_{x_2}, can be expressed by considering the discrete spatial frequency domain shown in Figure 3.10, as follows:

$$\begin{aligned} \Delta_{x_1} &= \frac{c}{2f_0 \Omega T_R} \\ \Delta_{x_2} &= \frac{c}{2\Delta_f} \end{aligned}$$

(3.57)

In order to avoid ambiguity, the relations in Equation (3.57) define the maximum sizes of the target. Any target bigger than the non-ambiguity region will be seen as folded within the image making target recognition more

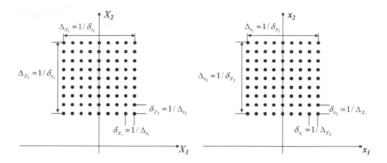

Figure 3.10 Non-ambiguity regions

difficult. It is worth pointing out that Δ_{x_1} depends also on the modulus of the effective rotation vector, Ω, which, as stated above, cannot be a priori predicted leading to a more difficult design of an ISAR system.

Moreover, considering Equation (3.24) some remarks about spatial resolutions can be made. Specifically, the range resolution depends on the transmitted signal bandwidth and can be directly controlled in the radar design process, since it can be considered as a known parameter. On the other hand, it is evident that the spatial resolution in the cross-range dimension depends on the total aspect angle variation, $\Delta\vartheta = \Omega T_{obs}$. This results in a parameter that is not controllable and cannot be selected during the radar design process, unless making some hypothesis about the target motion

3.4   CONCLUSIONS

The main concepts and theory behind ISAR imaging have been treated in this chapter. More specifically, the ISAR geometry and the received signal model have been defined. Then, mathematical details have been introduced in this chapter to support the main concepts behind ISAR technique and to provide the basis for implementing basic ISAR imaging algorithms. The concepts presented in this chapter mainly apply to monostatic active radar system. Nevertheless, the ISAR signal model formulation presented in this chapter is independent of the transmitted waveform; therefore, as it will be clarified in the following chapters, the same concepts apply to active and passive bistatic radar systems.

REFERENCES

1. F. Berizzi, M. Martorella, B. Haywood, E. Dalle Mese, and S. Bruscoli. A survey on ISAR autofocusing techniques. *Image Processing, 2004. ICIP '04. 2004 International Conference on*, 1:9–12, Oct. 2004.
2. F. Berizzi, E.D. Mese, M. Diani, and M. Martorella. High-resolution ISAR imaging of maneuvering targets by means of the range instantaneous doppler

technique: modeling and performance analysis. *Image Processing, IEEE Transactions on*, 10(12):1880–1890, Dec 2001.

3. G. Clifford Carter. Time delay estimation for passive sonar signal processing. *Acoustics, Speech and Signal Processing, IEEE Transactions on*, 29(3):463–470, Jun 1981.

4. B. Haywood and R. J. Evans. Motion compensation for ISAR imaging. *Proceedings of ASSPA 89*, pages 113–117, 1989.

5. C.V.J. Jakowatz, D.E. Wahl, P.H. Eichel, D.C. Ghiglia, and P.A. Thompson. *Spotlight-Mode Synthetic Aperture Radar: A Signal Processing Approach: A Signal Processing Approach*. Springer, 2011.

6. Steven M. Kay. *Fundamentals of Statistical Signal Processing: Estimation Theory*. Signal Processing. 1993.

7. M. Martorella. Novel approach for ISAR image cross-range scaling. *Aerospace and Electronic Systems, IEEE Transactions on*, 44(1):281–294, 2008.

8. M. Martorella, N. Acito, and F. Berizzi. Statistical CLEAN technique for ISAR imaging. *IEEE Tr. on Geoscience and Remote Sensing*, 45(11):3552–3560, 2007.

9. M. Martorella and F. Berizzi. Time windowing for highly focused ISAR image reconstruction. *Aerospace and Electronic Systems, IEEE Transactions on*, 41(3):992–1007, 2005.

10. M. Martorella, F. Berizzi, and B. Haywood. Contrast maximisation based technique for 2-D ISAR autofocusing. *Radar, Sonar and Navigation, IEE Proceedings*, 152(4):253–262, 2005.

11. R. Gray Parker and Ronald L. Rardin. *Discrete Optimization*. Academic Press Professional, Inc., San Diego, CA, 1988.

12. B.D. Steinberg. Radar imaging from a distorted array: The radio camera algorithm and experiments. *Antennas and Propagation, IEEE Transactions on*, 29(5):740–748, Sept. 1981.

13. J. Tsao and B.D. Steinberg. Reduction of sidelobe and speckle artifacts in microwave imaging: The CLEAN technique. *IEEE Tr. on Antennas and Propagation*, 36:543–556, 1988.

14. Yongmei Wei and Guoan Bi. Efficient analysis of time-varying multicomponent signals with modified LPTFT. *EURASIP J. Adv. Sig. Proc.*, 2005(8):1261–1268, 2005.

15. O.A. Yeste-Ojeda, J. Grajal, and G. Lopez-Risueno. Atomic decomposition for radar applications. *Aerospace and Electronic Systems, IEEE Transactions on*, 44(1):187 –200, Jan. 2008.

4 Bistatic ISAR

M. Martorella, D. Cataldo, S. Gelli, and S. Tomei

CONTENTS

Bistatic inverse synthetic aperture radar (B-ISAR) has progressed in the last decades pushed by technological advancements and by the need to overcome some limitations imposed by the monostatic radar geometry, as well as the need to image stealthy targets. Moreover, recent advances in bistatic radar imaging theory have paved the basis for the theoretical formulation of the multistatic ISAR imaging as well as passive radar imaging.

In order to obtain ISAR images with a fine Doppler resolution, the target is required to change its aspect angle with respect to the radar during the coherent processing interval (CPI). There are, however, cases in which, even if the target is moving with respect to the radar, an ISAR image cannot be obtained since the relative motion does not produce any change in the target aspect angle. As an example, when the target moves along the radar's line of sight (LoS), the target aspect angle does not change in time, and hence an ISAR image cannot be produced. Such a problem can be partially overcome by using a bistatic configuration. In fact, since the receiver position cannot be detected, a target cannot intentionally move along the bistatic LoS, namely the bisector of the bistatic angle (as it will be clarified later). Furthermore, in the event that multiple receivers are employed (multibistatic or multistatic systems), this problem can be fully overcome.

Radar imaging of stealthy targets can be enabled by using a bistatic configuration. In fact, stealthy targets are constructed to minimize the energy back-scattered toward the radar in order to make them almost invisible to the radar's eyes. This means that stealthiness usually refers only to monostatic radars, and hence the use of a bistatic radar may enable the target detection and enhance the radar image quality.

On the other hand, bistatic radar systems may suffer from distortions, due to the variation of the bistatic geometry during the CPI [15, 13, 14], and synchronization errors since transmitter and receiver are not usually co-located [11]. All these issues make the problem of forming bistatic ISAR images quite complex. In order to simplify the problem, the concept of bistatically equivalent monostatic (BEM) configuration has been recently introduced in literature, which allows for the bistatic geometry to be approximated with a monostatic one [13]. This greatly simplifies not only the geometrical understanding, but it also allows for monostatic ISAR processors to be used to form bistatic ISAR images. Such an approximation, however, is valid under some assumptions. The theoretical analysis of such constraints as well as the distortion effects due to changes in the bistatic angle during the observation time and synchronization errors can be found in [11, 15] and are reformulated in this chapter for the purpose of clarity and completeness.

The remainder of this chapter is organized as follows. section 4.1 presents the bistatic geometry and the signal modeling. The BEM concept is introduced in section 4.2. The B-ISAR image formation algorithm is detailed in section 4.3, where also the system point spread function (PSF) is derived. Furthermore, the bistatic geometry effects on the ISAR image formation are

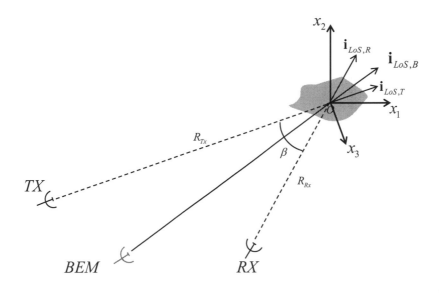

Figure 4.1 Bistatic geometry

analyzed. Further theoretical considerations, i.e., the analytical calculation of the effective rotation vector when considering a BEM configuration, and the distortions caused by the bistatic geometry are tackled in section 4.4. The synchronization error effects are finally addressed in section 4.5.

4.1 BISTATIC GEOMETRY AND SIGNAL MODELING

Referring to the mathematical notation introduced in chapter 1 and the geometry illustrated in figure 4.1, the received signal at the output of the matched filter, for a generic bistatic radar configuration, can be written in the frequency/slow time domain as follows:

$$S(f,n) = W(f,n) \int_V f'_B(\mathbf{x}) e^{-j2\pi f \tau(\mathbf{x},n)} d\mathbf{x} \qquad (4.1)$$

$$W(f,n) = K \cdot \mathrm{rect}\left(\frac{n}{N}\right) |S_T(f)|^2 = K \cdot \mathrm{rect}\left(\frac{n}{N}\right) \mathrm{rect}\left(\frac{f - f_0}{B}\right) \qquad (4.2)$$

where $f'_B(\mathbf{x})$ is the bistatic target reflectivity function defined in the volume V occupied by the target, f_0 and B are the central carrier frequency and the transmitted signal bandwidth, respectively, and $\tau(\mathbf{x}, n)$ is the delay time of a scatterer placed in $\mathbf{x} = (x_1, x_2, x_3)^T$, which is defined as follows:

$$\tau(\mathbf{x}, n) = \frac{R_{Tx}(\mathbf{x}, n) + R_{Rx}(\mathbf{x}, n)}{c} \tag{4.3}$$

where $R_{Tx}(\mathbf{x}, n)$ and $R_{Rx}(\mathbf{x}, n)$ are the distances between the scatterer and the transmitter and receiver, respectively, at the nth sweep.

As shown in figure 4.1, T_x is a Cartesian reference system embedded on the target and arbitrarily oriented. By assuming the radar target distances much greater than the size of the target (*straight iso-range* approximation), as in the monostatic case, the terms $R_{Tx}(\mathbf{x}, n)$ and $R_{Rx}(\mathbf{x}, n)$ can be approximated as follows:

$$R_{Tx}(\mathbf{x}, n) \simeq R_{Tx,0}(n) + \mathbf{x}^T \cdot \mathbf{i}_{LoS,T}(n) \tag{4.4}$$

$$R_{Rx}(\mathbf{x}, n) \simeq R_{Rx,0}(n) + \mathbf{x}^T \cdot \mathbf{i}_{LoS,R}(n) \tag{4.5}$$

where $R_{Tx,0}(n)$ and $R_{Rx,0}(n)$ are the distances between a reference point on the target and the transmitter and receiver, respectively, at the time instant nT_R, $\mathbf{i}_{LoS,T}(n)$ and $\mathbf{i}_{LoS,R}(n)$ are the column unit vectors that identify the LoS of the transmitter and receiver, respectively.

Then, by using Equation (4.4) and Equation (4.5), the delay time of a scatterer placed in $\mathbf{x}$ can be written as follows:

$$\begin{aligned}\tau(\mathbf{x}, n) &= \frac{2}{c} \left(\frac{R_{Tx,0}(n) + R_{Rx,0}(n)}{2} + \mathbf{x}^T \cdot \frac{\mathbf{i}_{LoS,T}(n) + \mathbf{i}_{LoS,R}(n)}{2} \right) \tag{4.6}\\ &= \frac{2}{c} \left(R_{B,0}(n) + K(n)\mathbf{x}^T \cdot \mathbf{i}_{LoS,B}(n) \right)\end{aligned}$$

where

$$R_{B,0}(n) = \frac{R_{Tx,0}(n) + R_{Rx,0}(n)}{2} \tag{4.7}$$

$$\mathbf{i}_{LoS,B}(n) = \frac{\mathbf{i}_{LoS,T}(n) + \mathbf{i}_{LoS,R}(n)}{|\mathbf{i}_{LoS,T}(n) + \mathbf{i}_{LoS,R}(n)|} \tag{4.8}$$

$$\begin{aligned}K(n) &= \left| \frac{\mathbf{i}_{LoS,T}(n) + \mathbf{i}_{LoS,R}(n)}{2} \right| \tag{4.9}\\ &= \cos\left(\frac{\arccos\left(\mathbf{i}_{LoS,T}^T(n) \cdot \mathbf{i}_{LoS,R}(n) \right)}{2} \right)\\ &= \cos\left(\frac{\beta(n)}{2} \right)\end{aligned}$$

In Equation (4.6) the term $R_{B,0}(n)$, which equals the mean value between the transmitter-target and the receiver-target distances, can be interpreted as a "bistatically equivalent monostatic radar-target distance", and the unit vector $\mathbf{i}_{LoS,B}(n)$ can be interpreted as a "bistatically equivalent monostatic LoS".

In chapter 3, the delay time in the monostatic case has been written as follows:

$$\tau_{\text{mono}}(\mathbf{x}, n) = \frac{2}{c} \left(R_0(n) + \mathbf{x}^T \cdot \mathbf{i}_{LoS}(n) \right) \tag{4.10}$$

Therefore, by comparing Equation (4.10) and Equation (4.6), it can be noted that the received signal model in the bistatic case is almost identical to the signal model in the monostatic case with the following differences:

- The function $f_B(\mathbf{x})$ in Equation (4.1) represents a bistatic reflectivity function which is different from the monostatic one $f(\mathbf{x})$;

- The bistatically equivalent monostatic distance $R_{B,0}(n)$;

- The bistatically equivalent monostatic LoS $\mathbf{i}_{LoS,B}(n)$;

- The presence of the distortion term $K(n)$, which depends on the bistatic angle β.

The presence of a bistatic reflectivity function instead of a monostatic one does not affect the image formation process, but only the final image interpretation. As for the monostatic case, the term $R_{B,0}(n)$ is responsible for range migration and defocusing effects and then must be suitably removed through a radial motion compensation step. However, $f_B(\mathbf{x})$ and $R_{B,0}(n)$ do not involve any substantial change in terms of ISAR image formation, as is demonstrated in the following sections.

Conversely, both $\mathbf{i}_{LoS,B}(n)$ and $K(n)$ deserve particular attention because they affect both the geometrical interpretation of the B-ISAR image and its quality.

4.2 BISTATICALLY EQUIVALENT MONOSTATIC APPROXIMATION

As shown in the previous section, the BEM approximation consists of reinterpreting the B-ISAR geometry in a convenient way by defining a BEM element as a monostatic antenna element, which acts as a virtual transceiver.

In particular, the BEM element is defined as a virtual transceiver located at a distance from the target equal to the mean value between the transmitter-target and the receiver-target distances along the direction of the bistatic angle bisector (see figure 4.1).

The distance between the BEM element and the target is $R_{B,0}(n)$ and the LoS unit vector is $\mathbf{i}_{LoS,B}(n)$.

Therefore, the main differences between a signal received by a monostatic radar located in the BEM element position and a signal received by the receiver of the bistatic pair are the reflectivity function $f(\mathbf{x})$ and the distortion term $K(n)$.

However, the term $K(n)$ does not affect the radial motion compensation. In the absence of synchronization errors, any parametric or non-parametric technique used with the monostatic ISAR can be applied to compensate the target radial motion with the desired accuracy [2]. More details about radial motion compensation and the image autofocus are given in section 4.5 where synchronization errors are also introduced.

4.3   BISTATIC ISAR IMAGE FORMATION AND POINT SPREAD FUNCTION

The term $K(n)$ carries information about the change in time of the bistatic geometry. This change during the CPI may significantly affect the final B-ISAR image. In this section, the B-ISAR image is analytically derived and the distortion introduced by the bistatic geometry is related to the bistatic angle variation. Two assumptions will be made that will allow for the use of the range-Doppler technique to reconstruct the ISAR image, namely:

1. Far field condition;
2. Relatively short integration time.

The second assumption means that the target rotation vector can be approximated as constant during the CPI and, moreover, that the Fourier domain of the received signal can be approximated as rectangular, thus avoiding the need for the use of polar reformatting. These assumptions are generally satisfied in typical ISAR scenarios where the resolutions required are not too fine.

When the target rotation vector can be considered constant, the received signal backscattered by a single ideal scatterer located at a generic point $\mathbf{x}_k = (x_{1_k}, x_{2_k}, x_{3_k})$ may be rewritten after motion compensation in the following way:

$$S_C(f, n) = W(f, n)\sigma_k e^{-j2\pi f\tau(\mathbf{x}_k, n)} \tag{4.11}$$

where σ_k is the reflectivity value of the ideal k^{th} scatterer (see Equation 3.15). Let us assume now that the Cartesian reference system embedded on the target T_x is chosen so that x_2 coincides with the range direction, namely $\mathbf{i}_{LoS,B}$ at the central time instant and x_3 coincides with the target effective rotation vector, $\mathbf{\Omega}$, namely:

$$\mathbf{i}_{x_2} \equiv \mathbf{i}_{LoS,B}(0) \tag{4.12}$$

$$\mathbf{i}_{x_3} \equiv \mathbf{i}_{\Omega} \tag{4.13}$$

$$\mathbf{i}_{x_1} = \mathbf{i}_{x_2} \times \mathbf{i}_{x_3} \tag{4.14}$$

where $\mathbf{i}_{x_1}$, $\mathbf{i}_{x_2}$ and $\mathbf{i}_{x_3}$ are the unit vectors which define the x_1, x_2 and x_3 axis, whereas $\mathbf{i}_{\Omega}$ is the effective rotation unit vector. Thus, the (x_1, x_2) 2D plane identifies the ISAR image projection plane (IPP) and the delay time can be written as follows:

$$\tau(\mathbf{x}_k, n) = \frac{2}{c} K(n) \mathbf{x}^T \cdot \mathbf{i}_{LoS,B}(n)$$

$$= \frac{2}{c} K(n) (x_{1_k} \sin(\Omega t_n) + x_{2_k} \cos(\Omega t_n)) \qquad (4.15)$$

where

$$t_n = n T_R \qquad (4.16)$$

and $(x_{1_k}, x_{2_k}, x_{3_k})$ are the coordinates of the considered generic scatterer on the target.

In Equation (4.15), Ω is the norm of the effective rotation vector $\mathbf{\Omega}$. In the monostatic case the effective rotation vector can be obtained from the target's rotation velocity vector $\mathbf{\Omega}_T$ by simply projecting it onto the 2D plane orthogonal to the LoS unit vector. In the bistatic case it is not so simple, but, as it will be clarified later, when the bistatic angle can be considered constant during the CPI, $\mathbf{\Omega}$ can be derived in the same way as in the monostatic case by considering the BEM approximation and the $\mathbf{i}_{LoS,B}$ unit vector. The calculation of the bistatic effective rotation vector is treated in section 4.4.

Under the previously mentioned assumptions, namely far field condition and short integration time, the bistatic angle changes are relatively small even when a target covers relatively large distances within the integration time. In this case, the bistatic angle can be well approximated with a first order Taylor-Maclaurin polynomial as follows:

$$\beta(n) \simeq \beta(0) + \dot{\beta}(0)t_n \qquad (4.17)$$

where $-T_{obs}/2 \leq t_n \leq T_{obs}/2$ and $\dot{\beta} = d\beta/dt$.

As a result, the term $K(n)$ can also be approximated with its first order Taylor-Maclaurin polynomial, and by using Equation (4.9) the following equation may be obtained:

$$K(n) \simeq K(0) + \dot{K}(0)t_n \qquad (4.18)$$

$$= \cos\left(\frac{\beta(0)}{2}\right) - \frac{\dot{\beta}(0)}{2} \sin\left(\frac{\beta(0)}{2}\right) t_n = K_0 + K_1 t_n$$

where the definition of K_0 and K_1 are intrinsically defined in Equation (4.18). Therefore, Equation (4.15) becomes:

$$\tau(\mathbf{x}_k, n) \simeq \frac{2}{c} (K_0 + K_1 t_n) (x_{1_k} \sin(\Omega t_n) + x_{2_k} \cos(\Omega t_n)) \qquad (4.19)$$

A short integration time means, however, small target aspect angle changes ($\theta \ll 1$), then even the $\cos(\Omega t_n)$ and $\sin(\Omega t_n)$ functions can be approximated by means of their first order Taylor-Maclaurin polynomial. As a consequence, Equation (4.19) can be written as follows:

$$\tau(\mathbf{x}_k, n) \simeq \frac{2}{c} (K_0 + K_1 t_n) (x_{1_k} \Omega t_n + x_{2_k}) \qquad (4.20)$$

Consequently, the received signal after motion compensation in Equation (4.11) can be written as follows:

$$S_C(f, n) = W(f, n) \cdot \sigma_k \cdot e^{-j\frac{4\pi f}{c}(K_0 + K_1 t_n)(x_{1_k} \Omega t_n + x_{2_k})} \qquad (4.21)$$

As in the monostatic case, the B-ISAR image formation is then achieved by means of the range-Doppler technique, which consists of a Fourier transform along both the frequency (range compression) and time (Doppler image formation) coordinates. Actually, the used signal model (see Equation (4.1)) is nothing else but the two-dimensional (2D) Fourier transform of the bistatic reflectivity function $f_B(\mathbf{x})$. Thus, the B-ISAR image formation consists of a 2D inverse Fourier transform (2D-IFT).

In the bistatic case, $K(n)$ carries information about the bistatic geometry changes over time. However, as it will be clearer later, what significantly affects the PSF of B-ISAR images is the term K_1. In the following subsections the B-ISAR PSF is analytically derived.

4.3.1   RANGE COMPRESSION

The range compression is obtained by transforming the signal in Equation (4.21) with respect to the frequency coordinate as follows:

$$
\begin{aligned}
S_{CR}(\tau, n) &= FT_f^{-1}\{S_C(f, n)\} \\
&= \int_{-\infty}^{\infty} \left(W(f, n) \cdot \sigma_k \cdot e^{-j2\pi f\left(\frac{2}{c}(K_0 + K_1 t_n)(x_{1_k} \Omega t_n + x_{2_k})\right)} \right) e^{j2\pi f\tau} \mathrm{d}f \\
&= \sigma_k \cdot \delta\left(\tau - \frac{2}{c}(K_0 + K_1 t_n)(x_{1_k}\Omega t_n + x_{2_k}) \right) \otimes_\tau \tilde{w}(\tau, n) \\
&= \sigma_k \cdot B \cdot \mathrm{rect}\left(\frac{n}{N}\right) \mathrm{sinc}\left\{ B\left(\tau - \frac{2}{c}(K_0 x_{2_k} + r_m(n)) \right) \right\} e^{j2\pi f_0 \tau} \\
&\quad \cdot \exp\left\{ -j\frac{4\pi f_0}{c}(K_0 + K_1 t_n)(x_{1_k}\Omega t_n + x_{2_k}) \right\} \qquad (4.22)
\end{aligned}
$$

where $\otimes_\tau$ is the convolution operator over the variable τ and

$$\tilde{w}(\tau, n) = FT_f^{-1}\{W(f, n)\} = \cdot \mathrm{rect}\left(\frac{n}{N}\right) B\mathrm{sinc}(B\tau) e^{j2\pi f_0 \tau} \qquad (4.23)$$

By observing Equation (4.22) it can be noted that two effects are induced by the bistatic geometry:

1. The range position x_{2_k} is scaled by a factor K_0, so the ideal scatterer is imaged in a new position $x'_{2_k} = K_0 x_{2_k}$;
2. A range migration induced by the bistatic angle variation during the CPI, which can be quantified with the term $r_m(n)$, namely

$$r_m(n) = K_0 x_{1,k}\Omega t_n + K_1 x_{2_k} t_n + K_1 x_{1_k}\Omega t_n^2 \qquad (4.24)$$

The first effect can be corrected by rescaling the range coordinate, but the second effect could be significantly detrimental. If range migration occurs, the position of one scatterer can be moved from one range cell to another during the integration time, thereby resulting in a distortion of the final image. To avoid range migration, the following inequality must be satisfied:

$$|r_m(n)| = \left|K_0 x_{1_k} \Omega t_n + K_1 x_{2_k} t_n + K_1 x_{1_k} \Omega t_n^2\right| < \frac{\delta_r}{2} = \frac{c}{4B} \qquad (4.25)$$

where δ_r is the radar range resolution.

It is worth pointing out that in the monostatic case $K_0 = 1$ and $K_1 = 0$ and the range migration term is equal to $|r_m(n)| = |x_{1_k} \Omega t_n|$. Since in a bistatic configuration $K_0 < 1$, the bistatic geometry may produce a higher range migration effect only when there is a significant bistatic angle variation component, i.e., large values of K_1. Consequently, since $|t_n| < T_{obs}/2$, it is important to satisfy the following constraint:

$$|K_1| T_{obs} x_{2_M} < \delta_r \qquad (4.26)$$

where x_{2_M} is the scatterer with maximum distance from the target focusing center.

By substituting the expression of K_1 in Equation (4.26) and by expressing the limitation with respect to the bistatic angle variation, the following relationship is obtained:

$$\left|\dot{\beta}(0)\right| < \frac{2\delta_r}{T_{obs} x_{2_M} \left|\sin\left(\frac{\beta(0)}{2}\right)\right|} \qquad (4.27)$$

In order to better understand the constraints of Equation (4.27), an example is considered. In figure 4.2, the curves (hyperbolas) of the bistatic angle variation ($\dot{\beta}(0)$) have been plotted against the bistatic angle ($\beta(0)$) for the parameters: $T_{obs} = 1$s, $x_{2_M} = 50$ m and for $\delta_r = [0.1, 0.4, 0.7, 1]$ m. From this figure it can be noticed that the tolerance about the bistatic angle change rate decreases in two cases: when the bistatic angle increases, and the range resolution decreases.

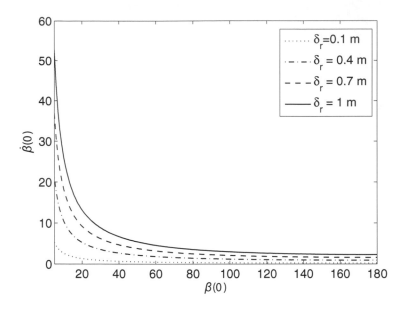

Figure 4.2 Bistatic deviation

When the constraint of Equation (4.27) is satisfied, $r_m(n)$ can be neglected and Equation (4.22) can be rewritten as follows:

$$
\begin{aligned}
S_{CR}(\tau, n) \;\simeq\; & \sigma_k \cdot B \cdot \operatorname{rect}\left(\frac{n}{N}\right) \operatorname{sinc}\left\{ B\left(\tau - \frac{2}{c} K_0 x_{2_k}\right)\right\} e^{j2\pi f_0 \tau} \\
& \cdot \exp\left\{-j\frac{4\pi f_0}{c}\left(K_0 + K_1 t_n\right)\left(x_{1_k}\Omega t_n + x_{2_k}\right)\right\} \\
=\; & \sigma_k \cdot B \cdot \operatorname{rect}\left(\frac{n}{N}\right) \operatorname{sinc}\left\{ B\left(\tau - \frac{2}{c} K_0 x_{2_k}\right)\right\} e^{j2\pi f_0\left(\tau - \frac{2}{c} K_0 x_{2_k}\right)} \\
& \cdot \exp\left\{-j2\pi f_0 \frac{2}{c}\left(\left(K_0 x_{1_k}\Omega + K_1 x_{2_k}\right) t_n + K_1 x_{1_k}\Omega t_n^2\right)\right\} \\
=\; & \sigma_k \cdot \delta\left(\tau - \frac{2}{c} K_0 x_{2_k}\right) \otimes_\tau \tilde{w}\left(\tau, t_n\right) \\
& \cdot \exp\left\{-j\frac{4\pi f_0}{c}\left(\left(K_0 x_{1_k}\Omega + K_1 x_{2_k}\right) t_n + K_1 x_{1_k}\Omega t_n^2\right)\right\} \quad (4.28)
\end{aligned}
$$

4.3.2 DOPPLER IMAGE FORMATION

Doppler image formation is achieved by applying an IFT to the range-compressed signal in Equation (4.28) along the slow-time variable. For the sake of convenience the discrete slow time variable t_n is considered as a continuous variable. The result is a complex image in the time-delay (range) and Doppler domain. Therefore, the final ISAR image can be calculated as follows [15, 14]:

$$I(\tau, \nu, x_{1_k}, x_{2_k}) = \tag{4.29}$$

$$= \sigma_k \, B \, \text{sinc}\left\{ B\left(\tau - \frac{2}{c} K_0 x_{2_k} \right) \right\} e^{j2\pi f_0 \left(\tau - \frac{2}{c} K_0 x_{2_k} \right)}.$$

$$\int_{-\infty}^{\infty} \text{rect}\left(\frac{t_n}{T_{obs}} \right) e^{\left\{ -j2\pi \frac{2f_0}{c} \left((K_0 x_{1_k} \Omega + K_1 x_{2_k}) t_n + K_1 x_{1_k} \Omega t_n^2 \right) \right\}} e^{j2\pi t_n \nu} dt_n$$

$$= \sigma_k \, \delta\left(\tau - \frac{2}{c} K_0 x_{2_k} \right) \delta\left(\nu - \frac{2f_0}{c} K_0 x_{1_k} \Omega - cr_s \right) \otimes_\tau \otimes_\nu w(\tau, \nu) \otimes_\nu D(\nu)$$

where $\otimes_\nu$ is the convolution operator over ν and

$$cr_s = \frac{2f_0}{c} K_1 x_{2_k} \tag{4.30}$$

$$D(\nu) = FT_{t_n}^{-1} \left\{ \exp\left\{ -j2\pi \left(\frac{2f_0}{c} K_1 x_{1_k} \Omega \right) t_n^2 \right\} \right\} \tag{4.31}$$

In particular, in Equation (4.29) cr_s is the shift term along the Doppler coordinate introduced by the bistatic angle change rate K_1, whereas $D(\nu)$ is a chirp-like distortion term. Moreover, the term $\frac{2}{c} K_0 x_{2_k}$ indicates that because of the bistatic geometry, the cross-range position x_{1_k} is scaled by a factor K_0. Therefore, the imaging PSF of the focusing point is represented by $w(\tau, \nu)$:

$$
\begin{aligned}
w(\tau, \nu) &= FT_{t_n}^{-1}\{\tilde{w}(\tau, t_n)\} \\
&= B \, \text{sinc}(B\tau) \, e^{j2\pi f_0 \tau} FT_{t_n}^{-1}\left\{ \text{rect}\left(\frac{t_n}{T_{obs}} \right) \right\} \\
&= B \cdot T_{obs} \, \text{sinc}(B\tau) \, \text{sinc}(T_{obs}\nu) \, e^{j2\pi f_0 \tau}
\end{aligned} \tag{4.32}
$$

As can be seen from Equation (4.31), the chirp rate depends on the position of the scatterer along the cross-range direction (x_{1_k}). The defocusing effect of a chirp signal has been largely studied in [1], where a new technique was proposed that eliminated such effects. By following [1] as a rule of thumb, when the chirp rate (see Equation (4.31)) satisfies the relationship in Equation (4.33) the defocusing effect can be neglected.

$$\left| \frac{2f_0}{c} K_1 x_{1_k} \Omega \right| < \frac{1}{T_{obs}^2} \tag{4.33}$$

By substituting the term K_1 as defined in Equation (4.18), into Equation (4.33), the following formula can be obtained which represents a rule for determining the maximum allowed bistatic angle variation to avoid defocusing effects:

$$\left|\dot{\beta}(0)\right| < \left|\frac{c}{f_0 T_{obs}^2 \Omega x_{1_M} \sin\left(\frac{\beta(0)}{2}\right)}\right| \qquad (4.34)$$

Under the assumption of a constant target's rotation vector and bistatic angle, β, cr_s and $D(\nu)$ can be considered negligible. Thus, the PSF of the B-ISAR image in Equation (4.32) is space invariant and $I(\tau, \nu, x_{1_k}, x_{2_k})$ can be approximated as follows:

$$I(\tau, \nu, x_{1_k}, x_{2_k}) \simeq \sigma_k \cdot \delta\left(\tau - \frac{2}{c}K_0 x_{2_k}\right) \delta\left(\nu - \frac{2f_0\Omega}{c}K_0 x_{1_k}\right) \otimes_\tau \otimes_\nu w(\tau, \nu)$$
$$(4.35)$$

As evident from Equation (4.35) and as mentioned before the B-ISAR image is a scaled version (by a factor K_0, see (Equation (4.35)) of the projected and filtered reflectivity function.

The space-invariant characteristic of the PSF guarantees a non-distorted B-ISAR image of the target, which is a desirable characteristic when target geometrical features are extracted for the purposes of classification and recognition.

In the next subsection, as an example, the maximum allowed target velocity to avoid distortions is derived in the worst operative scenario.

4.3.3   BISTATIC DISTORTION: WORST CASE

As it can be seen in Equation (4.30) and Equation (4.31), the bistatic distortion terms cr_s and $D(\nu)$ depend on the parameter $K_1 \propto \dot{\beta}(0)$ (see Equation (4.18)), i.e., the bistatic angle's variation speed. The worst operative scenario occurs when the target motion causes the greater bistatic angle variation within the observation time. The maximum bistatic angle variation occurs when the target is moving along the direction orthogonal to the baseline and crosses it at its center, as shown in figure 4.3.

By considering the velocity v positive when a target moves away from the radar baseline (B_L), the bistatic angle variation is equal to

$$\dot{\beta}(t) = \frac{B_L v}{\left[(R + vt)^2 + \left(\frac{B_L}{2}\right)^2\right]} \qquad (4.36)$$

Therefore, by applying Equation (4.34) and after some manipulation, the following inequality is obtained:

$$|v| < \left|\frac{c}{f_0 T_{obs}^2 \Omega x_{1_M} \sin\left(\frac{\beta(0)}{2}\right)}\right| \frac{1}{B_L} \left(R^2 + \left(\frac{B_L}{2}\right)^2\right) \qquad (4.37)$$

where B_L is the baseline and R is the distance between the target and the baseline center. Figure 4.4 shows the curves of the maximum velocity against

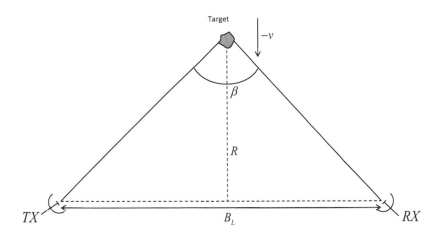

Figure 4.3 Bistatic distortion worst case geometry

the target distance R, for different values of the effective rotation vector's modulus Ω. Results in figure 4.4 have been obtained by using the following parameters: $f_0 = 10\,\text{GHz}$, $T_{obs} = 1\,\text{s}$, $x_{1M} = 50\,\text{m}$, $B_L = 5\,\text{km}$, $R \in [5 \div 20]\,\text{km}$ and $\Omega \in [0.01 \div 0.1]\,\text{rad/s}$.

4.4 BISTATIC IMAGE PROJECTION PLANE AND EFFECTIVE ROTATION VECTOR

The effective rotation vector and the B-ISAR image projection plane, when considering a BEM configuration, are derived mathematically in this section.

As it will be clearer later, the variation of the bistatic angle during the CPI may determine a variation of the angle between the range and cross-range unit vectors. As a result, such unit vectors may be not orthogonal to each other. In order to study this effect, let's now refer to the monostatic signal model after motion compensation, defined in Equation (3.1):

$$S_C(f, n) = W(f, n) \int_V f(\mathbf{x}) e^{-j\varphi(\mathbf{x}, f, n)} \, d\mathbf{x} \tag{4.38}$$

where

$$\varphi(\mathbf{x}, f, n) = \frac{4\pi f}{c} \mathbf{x}^T \cdot \mathbf{i}_{LoS}(n) \tag{4.39}$$

Let us now calculate the Doppler frequency term, which is obtained as the time derivative of the phase term in Equation (4.39). For the sake of simplicity,

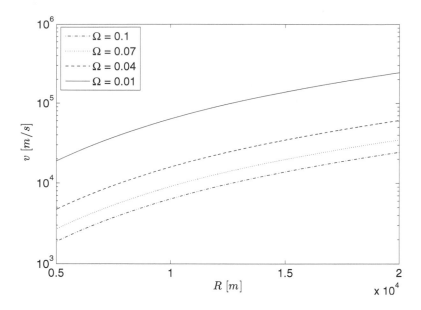

Figure 4.4   Maximum target speed in the worst case scenario

in the following the discrete "slow-time" variable n will be considered as a continuous variable, namely t. Therefore:

$$f_d(t) = \frac{1}{2\pi}\frac{d}{dt}\varphi(\mathbf{x}, f, t) = \frac{2f}{c}\frac{d}{dt}\left[\mathbf{x}^T \cdot \mathbf{i}_{LoS}(t)\right] \tag{4.40}$$

Referring to [10], $f_d(t)$ can be also expressed as follows:

$$f_d(t) = \frac{2f}{c}\left[\mathbf{\Omega} \times \mathbf{x}\right]^T \cdot \mathbf{i}_{LoS}(t) = \frac{2f}{c}\left[\mathbf{\Omega}_T^T\mathbf{L}(t)\mathbf{x}\right] = \frac{2f}{c}\left[\left(\mathbf{L}^T(t)\mathbf{\Omega}_T\right)^T\mathbf{x}\right] \tag{4.41}$$

where $\mathbf{\Omega}_T$ is the total target rotation (column) vector, $\mathbf{\Omega} = \mathbf{i}_{LoS} \times (\mathbf{\Omega}_T \times \mathbf{i}_{LoS})$ is the effective rotation vector and $\mathbf{L}(t)$ is the following matrix:

$$\mathbf{L}(t) = \begin{bmatrix} 0 & \sin(\theta_e(t)) & -\sin(\theta_a(t))\cos(\theta_e(t)) \\ -\sin(\theta_e(t)) & 0 & \cos(\theta_a(t))\cos(\theta_e(t)) \\ \sin(\theta_a(t))\cos(\theta_e(t)) & -\cos(\theta_a(t))\cos(\theta_e(t)) & 0 \end{bmatrix}$$

$$\tag{4.42}$$

In particular, in Equation (4.42) $\theta_a(t)$ and $\theta_e(t)$ are the azimuth and elevation angles which identify the LoS direction with respect to the reference system embedded on the target. Both $\mathbf{\Omega}$ and $\mathbf{\Omega}_T$ are assumed to be constant during the CPI.

It can be demonstrated that $\mathbf{L}^T(t)\mathbf{\Omega}_T = \mathbf{i}_{LoS}(t) \times \mathbf{\Omega}_T$. Then Equation (4.41) can be written as follows:

$$f_d(t) = \frac{2f}{c}\left[(\mathbf{i}_{LoS}(t) \times \mathbf{\Omega}_T)^T \cdot \mathbf{x}\right] \tag{4.43}$$

Therefore, by combining Equation (4.40) and Equation (4.43), the following result can be obtained:

$$\frac{\mathrm{d}}{\mathrm{d}t}\left[\mathbf{x}^T \cdot \mathbf{i}_{LoS}(t)\right] = (\mathbf{i}_{LoS}(t) \times \mathbf{\Omega}_T)^T \cdot \mathbf{x} \tag{4.44}$$

Equation (4.44) will be used for the computation of the bistatic effective rotation vector $\mathbf{\Omega}_B$.

4.4.1 BISTATIC EFFECTIVE ROTATION VECTOR

To derive the bistatic effective rotation vector $\mathbf{\Omega}_B$, the first step consists of calculating the Doppler frequency through the derivative function of the phase term in Equation (4.6) without the term $R_{B,0}(n)$. Also in this case the continuous time variable t is considered instead of the discrete one, n. In this case, the Doppler frequency can be written as:

$$\begin{aligned} f_d(t) &= \frac{1}{2\pi}\frac{\mathrm{d}}{\mathrm{d}t}\varphi(t) = \frac{2f}{c}\frac{\mathrm{d}}{\mathrm{d}t}\left[\frac{\mathbf{x}^T \cdot \mathbf{i}_{LoS,T}(t) + \mathbf{x}^T \cdot \mathbf{i}_{LoS,R}(t)}{2}\right] \\ &= \frac{2f}{c}\frac{\mathrm{d}}{\mathrm{d}t}\left[K(t)\mathbf{x}^T \cdot \mathbf{i}_{LoS,B}(t)\right] \end{aligned} \tag{4.45}$$

Then, the following equation holds:

$$\frac{1}{2}\frac{\mathrm{d}}{\mathrm{d}t}\left[\mathbf{x}^T \cdot \mathbf{i}_{LoS,T}(t) + \mathbf{x}^T \cdot \mathbf{i}_{LoS,R}(t)\right] = \frac{\mathrm{d}}{\mathrm{d}t}\left[K(t)\mathbf{x}^T \cdot \mathbf{i}_{LoS,B}(t)\right] \tag{4.46}$$

Then, by using the results in Equation (4.44), Equation (4.47) can be derived:

$$\begin{aligned} \frac{1}{2}&\left[(\mathbf{i}_{LoS,T}(t) \times \mathbf{\Omega}_{Tx})^T \cdot \mathbf{x} + (\mathbf{i}_{LoS,R}(t) \times \mathbf{\Omega}_{Rx})^T \cdot \mathbf{x}\right] = \\ &\dot{K}(t)\mathbf{x}^T \cdot \mathbf{i}_{LoS,B}(t) + K(t)(\mathbf{i}_{LoS,B}(t) \times \mathbf{\Omega}_{BT})^T \cdot \mathbf{x} \end{aligned} \tag{4.47}$$

where $\mathbf{\Omega}_{Tx}$, $\mathbf{\Omega}_{Rx}$ and $\mathbf{\Omega}_{BT}$ are the total rotation column vectors with respect to the transmitter, the receiver and the BEM, respectively, and $\dot{K}(t) = \frac{\mathrm{d}K(t)}{\mathrm{d}t}$.

After some mathematical manipulation, Equation (4.48) can be derived:

$$\begin{aligned} \frac{1}{2}&(\mathbf{i}_{LoS,T}(t) \times \mathbf{\Omega}_{Tx} + \mathbf{i}_{LoS,R}(t) \times \mathbf{\Omega}_{Rx}) = \\ &\dot{K}(t)\mathbf{i}_{LoS,B}(t) + K(t)(\mathbf{i}_{LoS,B}(t) \times \mathbf{\Omega}_{BT}) \end{aligned} \tag{4.48}$$

By knowing that the effective rotation vector can be found as $\boldsymbol{\Omega} = (\mathbf{i}_{LoS} \times \boldsymbol{\Omega}_T) \times \mathbf{i}_{LoS}$ and that the radar LoS, $\mathbf{i}_{LoS}$, the effective rotation vector, $\boldsymbol{\Omega}$ and the cross-range unit vector $\mathbf{i}_{CR}$ form a right-handed coordinate system, the vector $\tilde{\mathbf{i}}_{CR}$ can be defined as follows:

$$\tilde{\mathbf{i}}_{CR}(t) = \mathbf{i}_{LoS,B}(t) \times \boldsymbol{\Omega}_{BT} \tag{4.49}$$

$$= \frac{(\mathbf{i}_{LoS,T}(t) \times \boldsymbol{\Omega}_{Tx}) + (\mathbf{i}_{LoS,R}(t) \times \boldsymbol{\Omega}_{Rx}) - \left(2\dot{K}(t)\,\mathbf{i}_{LoS,B}(t)\right)}{2K(t)}$$

However, it is important to note that $\tilde{\mathbf{i}}_{CR}(t)$ is not a unit vector, but instead it can be rewritten as

$$\tilde{\mathbf{i}}_{CR}(t) = |\tilde{\mathbf{i}}_{CR}(t)|\hat{\mathbf{i}}_{CR} \tag{4.50}$$

The bistatic effective rotation vector $\boldsymbol{\Omega}_B$ can be calculated as follows:

$$\boldsymbol{\Omega}_B = \tilde{\mathbf{i}}_{CR}(t) \times \mathbf{i}_{LoS,B}(t) \tag{4.51}$$

$$= \frac{(\mathbf{i}_{LoS,T}(t) \times \boldsymbol{\Omega}_{Tx}) + (\mathbf{i}_{LoS,R}(t) \times \boldsymbol{\Omega}_{Rx}) - \left(2\dot{K}(t)\mathbf{i}_{LoS,B}(t)\right)}{2K(t)} \times \mathbf{i}_{LoS,B}(t)$$

$$= \frac{(\mathbf{i}_{LoS,T}(t) \times \boldsymbol{\Omega}_{Tx}) + (\mathbf{i}_{LoS,R}(t) \times \boldsymbol{\Omega}_{Rx})}{2K(t)} \times \mathbf{i}_{LoS,B}(t)$$

It is worth pointing out that as the baseline decreases to zero, the bistatic configuration approaches to the monostatic one. As a consequence,

$$\lim_{B_L \to 0} \boldsymbol{\Omega}_B = (\mathbf{i}_{LoS}(t) \times \boldsymbol{\Omega}_T) \times \mathbf{i}_{LoS}(t) \tag{4.52}$$

In other words, the monostatic configuration can be seen as a particular case of the bistatic one, in which:

$$\mathbf{i}_{LoS,T}(t) = \mathbf{i}_{LoS,R}(t) = \mathbf{i}_{LoS,B}(t) = \mathbf{i}_{LoS}(t) \tag{4.53}$$

$$\boldsymbol{\Omega}_{Tx} = \boldsymbol{\Omega}_{Rx} = \boldsymbol{\Omega}_T \tag{4.54}$$

and $K(t) = 1$. Finally, it is worth saying that the bistatic image projection plane (B-IPP) is defined as the plane orthogonal to $\boldsymbol{\Omega}_B$.

4.4.2   CROSS-RANGE UNIT VECTOR: FURTHER ANALYSIS

A deeper analysis of the cross-range unit vector in the bistatic case is addressed in this subsection.

As is well known, in the monostatic case the Doppler frequency term is linked to the cross-range unit vector as follows:

$$f_d(t) = \frac{2f}{c}\Omega \mathbf{i}_{CR}^T \cdot \mathbf{x} \tag{4.55}$$

where $\mathbf{i}_{CR}$ is the cross-range unit vector and $\Omega = |\mathbf{\Omega}|$. However, the Doppler frequency can be also written by means of Equation (4.41) as follows:

$$f_d(t) = \frac{2f}{c} [\mathbf{\Omega} \times \mathbf{x}]^T \cdot \mathbf{i}_{LoS}(t) = \frac{2f}{c} [\mathbf{i}_{LoS}(t) \times \mathbf{\Omega}_T]^T \cdot \mathbf{x} = \frac{2f}{c} [\mathbf{i}_{LoS}(t) \times \mathbf{\Omega}]^T \cdot \mathbf{x}$$

(4.56)

In the bistatic case, the Doppler frequency can be obtained by means of Equations (4.44) and (4.45):

$$f_d(t) = \frac{2f}{c} \frac{(\mathbf{i}_{LoS,T}(t) \times \mathbf{\Omega}_{Tx})^T + (\mathbf{i}_{LoS,R}(t) \times \mathbf{\Omega}_{Rx})^T}{2} \cdot \mathbf{x}$$

(4.57)

Therefore, by comparing Equation (4.57) with Equation (4.55), the cross-range unit vector for a bistatic radar system can be written as follows:

$$\mathbf{i}_{CR,B}(t) \doteq \frac{(\mathbf{i}_{LoS,T}(t) \times \mathbf{\Omega}_{Tx}) + (\mathbf{i}_{LoS,R}(t) \times \mathbf{\Omega}_{Rx})}{|(\mathbf{i}_{LoS,T}(t) \times \mathbf{\Omega}_{Tx}) + (\mathbf{i}_{LoS,R}(t) \times \mathbf{\Omega}_{Rx})|}$$

(4.58)

The vector in Equation (4.49), namely $\tilde{\mathbf{i}}_{CR}(t)$, has been obtained so as to be orthogonal to both $\mathbf{i}_{LoS,B}$ and $\mathbf{\Omega}_{BT}$, then it is alway orthogonal to the BEM LoS:

$$\tilde{\mathbf{i}}_{CR}(t) \perp \mathbf{i}_{LoS,B}(t)$$

(4.59)

Therefore, by manipulating Equation (4.49) and by using Equation (4.58), $\tilde{\mathbf{i}}_{CR}(t)$ can be rewritten as a function of $\mathbf{i}_{CR,B}(t)$:

$$\begin{aligned}
\tilde{\mathbf{i}}_{CR}(t) &= \frac{(\mathbf{i}_{LoS,T}(t) \times \mathbf{\Omega}_{Tx}) + (\mathbf{i}_{LoS,R}(t) \times \mathbf{\Omega}_{Rx})}{2K(t)} - \frac{\dot{K}(t)}{K(t)} \mathbf{i}_{LoS,B}(t) \quad (4.60) \\
&= \frac{|(\mathbf{i}_{LoS,T}(t) \times \mathbf{\Omega}_{Tx}) + (\mathbf{i}_{LoS,R}(t) \times \mathbf{\Omega}_{Rx})|}{2K(t)} \mathbf{i}_{CR,B}(t) - \frac{\dot{K}(t)}{K(t)} \mathbf{i}_{LoS,B}(t) \\
&= [\alpha_1 \mathbf{i}_{CR,B}(t) + \alpha_2 \mathbf{i}_{LoS,B}(t)] \perp \mathbf{i}_{LoS,B}(t)
\end{aligned}$$

where α_1 and α_2 are scalar quantities. This result means that $\mathbf{i}_{CR,B}(t)$ and $\mathbf{i}_{LoS,B}(t)$ are not orthogonal.

Then, $\tilde{\mathbf{i}}_{CR}(t)$ can be seen as an "orthogonalized" non-unit cross-range vector. In the case of small bistatic angle variations, the $K(t)$ term can be approximated as follows:

$$K(t) \simeq K_0 \quad ; \quad \dot{K}(t) \simeq K_1$$

(4.61)

Then, Equation (4.60) can be rewritten as follows:

$$\tilde{\mathbf{i}}_{CR}(t) \simeq \frac{|(\mathbf{i}_{LoS,T}(t) \times \mathbf{\Omega}_{Tx}) + (\mathbf{i}_{LoS,R}(t) \times \mathbf{\Omega}_{Rx})|}{2K_0} \mathbf{i}_{CR,B}(t) - \frac{K_1}{K_0} \mathbf{i}_{LoS,B}(t)$$

(4.62)

and the bistatic cross-range unit vector becomes:

$$\mathbf{i}_{CR,B}(t) \simeq \frac{2K_0}{|(\mathbf{i}_{LoS,T}(t) \times \boldsymbol{\Omega}_{Tx}) + (\mathbf{i}_{LoS,R}(t) \times \boldsymbol{\Omega}_{Rx})|} \tilde{\mathbf{i}}_{CR}(t) \quad (4.63)$$

$$+ \frac{2K_1}{|(\mathbf{i}_{LoS,T}(t) \times \boldsymbol{\Omega}_{Tx}) + (\mathbf{i}_{LoS,R}(t) \times \boldsymbol{\Omega}_{Rx})|} \mathbf{i}_{LoS,B}(t)$$

Equation (4.63) shows that the bistatic cross-range unit vector is composed of two components: the former, orthogonal to the LoS unit vector and proportional to K_0, the latter, parallel to $\mathbf{i}_{LoS,B}(t)$ and proportional to K_1.

However, since both $\tilde{\mathbf{i}}_{CR}(t)$ and $\mathbf{i}_{LoS,B}(t)$ are orthogonal to the bistatic effective rotation vector $\boldsymbol{\Omega}_B$ (see Equation (4.51)), even if the parameter K_1 makes the range and cross-range unit vectors not orthogonal, it does not affect the B-IPP. Thus, the image projection plane does not depend on the bistatic angle variation.

4.4.3   BISTATIC LINEAR DISTORTION: FURTHER ANALYSIS

In the previous subsection the non-orthogonality between the range and cross-range unit vector in a bistatic radar system has been demonstrated and analyzed. This means that when the target's bistatic three-dimensional reflectivity function $f_B(\mathbf{x})$ is projected onto the B-IPP, the non-orthogonality between the cross-range and range unit vectors determines a further scaling effect along the cross-range direction. The scaling factor is equal to the cosine of the angle between $\tilde{\mathbf{i}}_{CR}(t)$ and $\mathbf{i}_{CR,B}(t)$.

Moreover, by comparing Equation (4.58) with Equation (4.63), the following equation can be written:

$$\frac{(\mathbf{i}_{LoS,T}(t) \times \boldsymbol{\Omega}_{Tx}) + (\mathbf{i}_{LoS,R}(t) \times \boldsymbol{\Omega}_{Rx})}{2} = K_0 \tilde{\mathbf{i}}_{CR}(t) + K_1 \mathbf{i}_{LoS,B}(t) \quad (4.64)$$

Therefore, by merging Equations (4.44), (4.45) and (4.64) we obtain:

$$\begin{aligned} f_d(t) &= \frac{2f}{c} \frac{\mathrm{d}}{\mathrm{d}t} \left[\frac{\mathbf{x}^T \cdot \mathbf{i}_{LoS,T}(t) + \mathbf{x}^T \cdot \mathbf{i}_{LoS,R}(t)}{2} \right] \\ &= \frac{2f}{c} \left[\frac{(\mathbf{i}_{LoS,T}(t) \times \boldsymbol{\Omega}_{Tx})^T + (\mathbf{i}_{LoS,R}(t) \times \boldsymbol{\Omega}_{Rx})^T}{2} \cdot \mathbf{x} \right] \\ &= \frac{2f}{c} \left[K_0(\tilde{\mathbf{i}}_{CR}^T(t) \cdot \mathbf{x}) + K_1(\mathbf{i}_{LoS,B}(t)^T \cdot \mathbf{x}) \right] \\ &= \frac{2f}{c} \left[K_0 \left| \tilde{\mathbf{i}}_{CR}(t) \right| (\hat{\mathbf{i}}_{CR}^T(t) \cdot \mathbf{x}) + K_1(\mathbf{i}_{LoS,B}^T(t) \cdot \mathbf{x}) \right] \quad (4.65) \end{aligned}$$

where $\hat{\mathbf{i}}_{CR}(t)$ is the "orthogonalized" cross-range unit vector:

$$\hat{\mathbf{i}}_{CR}(t) \doteq \frac{\tilde{\mathbf{i}}_{CR}(t)}{\left| \tilde{\mathbf{i}}_{CR}(t) \right|} \quad (4.66)$$

Since $\boldsymbol{\Omega}_B = \tilde{\mathbf{i}}_{CR}(t) \times \mathbf{i}_{LoS,B}(t)$ and $\tilde{\mathbf{i}}_{CR}(t)$ and $\mathbf{i}_{LoS,B}(t)$ are perpendicular to each other, the following equation holds:

$$|\boldsymbol{\Omega}_B| = \left|\tilde{\mathbf{i}}_{CR}(t) \times \mathbf{i}_{LoS,B}(t)\right| = \left|\tilde{\mathbf{i}}_{CR}(t)\right| \cdot |\mathbf{i}_{LoS,B}(t)| = \left|\tilde{\mathbf{i}}_{CR}(t)\right|$$

Thus the Doppler frequency:

$$f_d(t) = \frac{2f}{c}\left[K_0\Omega_B(\hat{\mathbf{i}}_{CR}^T(t) \cdot \mathbf{x}) + K_1(\mathbf{i}_{LoS,B}^T(t) \cdot \mathbf{x})\right] = \tilde{f}_d(t) + \Delta f_d(t) \quad (4.67)$$

where

$$\Omega_B \doteq |\boldsymbol{\Omega}_B| \quad (4.68)$$

$$\tilde{f}_d(t) \doteq \frac{2f}{c}\Omega_B K_0(\hat{\mathbf{i}}_{CR}^T(t) \cdot \mathbf{x}) \quad (4.69)$$

$$\Delta f_d(t) \doteq \frac{2f}{c}K_1(\mathbf{i}_{LoS,B}^T(t) \cdot \mathbf{x}) \quad (4.70)$$

The result in Equation (4.67) means that the Doppler frequency generated by a scatterer located in $\mathbf{x}$ depends on the position along both the range and the "orthogonalized" cross-range directions. Therefore, since the reference system T_x has been chosen to have

$$\mathbf{i}_{x_3} \equiv \mathbf{i}_\Omega \doteq \frac{\boldsymbol{\Omega}_B}{\Omega_B} \quad (4.71)$$

(see Equation (4.13)) and

$$\mathbf{x}^T \cdot \mathbf{i}_{LoS,B}(t) = [x_1 \sin(\Omega_B t) + x_2 \cos(\Omega_B t)] \quad (4.72)$$

the Doppler shift $\Delta f_d(t)$ can be written as follows:

$$\begin{aligned} \Delta f_d(t) &= \frac{2f}{c}K_1[x_1\sin(\Omega_B t) + x_2\cos(\Omega_B t)] \quad (4.73)\\ &\simeq \frac{2f}{c}K_1 x_1 \Omega_B t + \frac{2f}{c}K_1 x_2 \end{aligned}$$

It is possible to note that the Doppler shift consists of two Doppler terms. Specifically,

$$\Delta f_{d,2}(t) \doteq \frac{2f}{c}K_1 x_2 \quad (4.74)$$

which coincides with the shift term cr_s defined in Equation (4.30) and

$$\Delta f_{d,1}(t) \doteq \frac{2f}{c}K_1 x_1 \Omega_B t \quad (4.75)$$

which contributes to a chirp-like distortion term (already introduced in Equation (4.31)) which is defined as follows:

$$D(\nu) = FT_t\left\{e^{-j2\pi\Delta f_{d,1}(t)t}\right\} \quad (4.76)$$

Furthermore, by comparing Equation (4.55) with Equation (4.69) it appears that the bistatic geometry introduces a scaling effect equivalent to K_0 along the cross-range direction, as also demonstrated in section 4.3.2 by other means.

In summary, the bistatically equivalent monostatic approximation can be used if the bistatic angle does not vary too fast, i.e., when K_1 is negligible. In this case, the final ISAR image will be the same as that obtained with a monostatic radar (considering the effective rotation vector Ω_B in Equation (4.51)), with only a scaling effect equivalent to K_0 (in both range and cross-range directions).

In order to have negligible linear distortion effects, the maximum value of $|\Delta f_d(t)|$ must be smaller than the Doppler resolution:

$$|\Delta f_d(t)| < \frac{1}{T_{obs}} \qquad (4.77)$$

Equation (4.77) can be rewritten as a function of the K_1 term as follows:

$$|K_1| < \frac{c}{2 f T_{obs} x_{2_M}} \qquad (4.78)$$

where x_{2_M} is the distance along the range direction between the target focusing center and the furthest scatterer, or equivalently the maximum value of $|\mathbf{i}_{LoS,B}(t) \cdot \mathbf{x}|$.

4.5 B-ISAR IN THE PRESENCE OF PHASE SYNCHRONIZATION ERRORS

In the previous sections, the B-ISAR has been investigated paying attention to the effects of the bistatic geometry on the system PSF. Thus, the distortion induced by the bistatic geometry has been analyzed with respect to the monostatic case.

Bistatic radar systems, however, suffer also from synchronization errors since the transmitter and receiver are not colocated. More specifically, synchronization errors can be classified depending on their nature: spatial, temporal, and phase [11].

With spatial synchronization, we refer to the synchronization of the transmitting and receiving antenna beams, which should be directed toward the same physical location [18]. Spatial synchronization is needed because antennas with high gain have a narrow beam pattern, which results in a relatively small illuminated area. In a bistatic system, therefore, the transmitting and the receiving antenna beams must be directed toward the same area to maximize the received signal-to-noise ratio (SNR) and consequently the SNR in the formed ISAR image.

Temporal synchronization allows for correct target ranging and time gating for the selected data subset to be processed in order to form an ISAR image and requires a common time reference that must be used for measuring delays. In software-defined radar systems, a finer time synchronization is also needed because the same time reference is used to synchronize analog-to-digital converters (ADC) and field programmable gate arrays (FPGAs) for signal processing, including signal down-conversion, demodulation and filtering.

Phase synchronization requires coherency between the two local radio frequency (RF) oscillators that operate in the transmitter and receiver [17] for signal down-conversion in standard radar RF front ends. In a monostatic system, synchronization is automatically obtained because a single antenna is used for both transmitting and receiving, a single clock is used as time reference, and a single local oscillator is used for generating, modulating, and demodulating the transmitted and received signals. On the contrary, in bistatic systems two separate oscillators are used to up-convert (transmitting station) and down-convert (receiving station) the RF signal and coherence must be guaranteed by synchronizing the two oscillators. The problem of synchronization has been addressed in the B-SAR literature [8, 9]. Nevertheless, given the differences that exist between SAR and ISAR, a detailed analysis is also required for the ISAR case. This latter issue is in fact dominant in radar systems that require a high level of coherency.

In this section, the robustness of standard radial motion compensation techniques and an RD-based monostatic ISAR processor are analyzed when applied to bistatic geometries in the presence of such phase synchronization errors. In particular, the B-ISAR point spread function (PSF) is analytically obtained. Parametric constraints that define the limits of the usability of a monostatic RD processor are then derived.

4.5.1   SIGNAL MODELING IN THE PRESENCE OF SYNCHRONIZATION ERRORS

The geometry of the problem is illustrated in figure 4.1, whereas the used signal model has been defined in Equation (4.1):

$$S(f,t) = W(f,t) \int_V f'_B(\mathbf{x}) e^{-j\varphi(\mathbf{x},f,t)} \mathrm{d}\mathbf{x} \qquad (4.79)$$

For convenience, in this case both frequency and time domains are assumed continuous.

The phase term $(\varphi(\mathbf{x}, f, t) = 2\pi f \tau(\mathbf{x}, t))$ in Equation (4.1) was considered to be only dependent on the radar-target configuration. In this section, such a result is extended to a more realistic scenario where phase synchronization errors are present. In order to account for them, synchronization errors are modeled as a frequency jitter (FJ):

$$\Delta f(t) = \eta_0 + \eta_1 t + \delta F(t) \qquad (4.80)$$

where η_0 is a frequency offset, $\eta_1 t$ is a linear frequency drift, which both usually depend on the oscillator temperature, and $\delta F(t)$ is a random FJ that accounts for both the transmitter and receiver local oscillators (noise-like).

Oscillator synchronization errors introduce an effect on the phase term in the received signal. Such an effect can be analyzed separately by considering the phase term associated with the target radial motion compensation and the phase term associated with the target rotational motion.

Therefore, the phase terms relative to the bistatic configuration can be written as follows (see Equations (4.3) and (4.6)):

$$\varphi(\mathbf{x}, f, t) = 2\pi \left[f + \Delta f(t) \right] \frac{R_{Tx}(\mathbf{x}, n) + R_{Rx}(\mathbf{x}, n)}{c} \tag{4.81}$$

$$= 2\pi \left[f + \Delta f(t) \right] \frac{2 \left(R_{B,0}(t) + K(t) \, \mathbf{x}^T \cdot \mathbf{i}_{LoS,B}(t) \right)}{c}$$

$$= \varphi_0(f, t) + \varphi_x(\mathbf{x}, f, t)$$

where

$$\varphi_0(f, t) = 2\pi \left[f + \Delta f(t) \right] \frac{2 R_{B,0}(t)}{c} \tag{4.82}$$

is the reference phase term associated with the spatial position of the target focusing center and

$$\varphi_x(\mathbf{x}, f, t) = 2\pi \left[f + \Delta f(t) \right] \frac{2 K(t) \, \mathbf{x}^T \cdot \mathbf{i}_{LoS,B}(t)}{c} \tag{4.83}$$

is the phase term from a scattering point positioned at $\mathbf{x}$.

4.5.2 RADIAL MOTION COMPENSATION IN THE PRESENCE OF SYNCHRONIZATION ERRORS

Often parametric ISAR image autofocusing algorithms are employed because they provide more accurate solutions to the radial motion compensation problem [2]. The effects of bistatic synchronization errors are analyzed here for a specific algorithm, namely the image contrast based autofocus (ICBA) [12], although the conclusions can be extended to any parametric and non-parametric autofocus technique. In order to use such parametric algorithms, a bistatically equivalent radial motion phase term model must be derived. A quadratic polynomial phase term is generally used as an approximation for the radial motion in the monostatic case, as detailed in Equation (4.84):

$$R_{B,0}(t) = \frac{R_{Tx,0}(t) + R_{Rx,0}(t)}{2} \simeq \alpha_0 + \alpha_1 t + \alpha_2 t^2 \tag{4.84}$$

The validity of the monostatic approximation in the bistatic case can be simply demonstrated. In fact, the bistatic phase history associated with an arbitrary scatterer on the target can be seen as the semi-sum of two monostatic phase histories. Therefore, if the monostatic phase histories can be approximated with n^{th} order polynomials, then the semi-sum of them can be approximated with a polynomial of the same order. Therefore, by taking into account the approximation in Equation (4.84), the phase relative to the target's radial motion (to be compensated) can be obtained by substituting Equations (4.84) and (4.80) into Equation (4.82):

$$\varphi_0(f,t) = \frac{4\pi \left[f + \eta_0 + \eta_1 t + \delta F(t) \right]}{c} \left[\alpha_0 + \alpha_1 t + \alpha_2 t^2 \right] \simeq$$

$$\frac{4\pi}{c} \left\{ (f + \eta_0) \alpha_0 + \left[(f + \eta_0) \alpha_1 + \eta_1 \alpha_0 \right] t + \left[(f + \eta_0) \alpha_2 + \eta_1 \alpha_1 \right] t^2 + \eta_1 \alpha_2 t^3 \right\}$$

(4.85)

It is worth noting that the term $\delta F(t)$ is usually much smaller than the frequency offset and linear drift and therefore it has been neglected in Equation (4.85). It should also be pointed out that the resultant phase noise variance in a B-ISAR system would be the sum of the transmitter and receiver oscillator phase noise variances (independence of the oscillators), and therefore it is usually slightly larger than in the monostatic case. Furthermore, $\varphi_0(f,t)$ can be seen as a consequence of virtual radial motion, which differs from the actual radial motion because of the presence of synchronization errors. It is shown that the presence of a phase offset and a phase drift do not significantly affect the estimation of the bistatically equivalent radial motion parameters.

A three-parameter estimation problem can be stated by rewriting Equation (4.85) as follows:

$$\varphi_0(f,t) = \frac{4\pi}{c} \left(\xi_0 + \xi_1 t + \xi_2 t^2 + \xi_3 t^3 \right) \qquad (4.86)$$

where

$$\xi_0 = (f + \eta_0) \alpha_0 \qquad (4.87)$$

$$\xi_1 = (f + \eta_0) \alpha_1 + \eta_1 \alpha_0 \qquad (4.88)$$

$$\xi_2 = (f + \eta_0) \alpha_2 + \eta_1 \alpha_1 \qquad (4.89)$$

$$\xi_3 = \eta_1 \alpha_2 \qquad (4.90)$$

As demonstrated in chapter 3, the term ξ_0 can be ignored when applying parametric autofocusing techniques since it only induces a range shift in an ISAR image. Equation (4.86) can be rewritten as follows, by neglecting ξ_0:

$$\varphi_0(f,t) = \frac{4\pi f}{c} \left(\xi_1' t + \xi_2' t^2 + \xi_3' t^3 \right) \qquad (4.91)$$

where

$$\xi_1' = \left(1 + \frac{\eta_0}{f} \right) \alpha_1 + \frac{\eta_1 \alpha_0}{f} \qquad (4.92)$$

$$\xi_2' = \left(1 + \frac{\eta_0}{f} \right) \alpha_2 + \frac{\eta_1 \alpha_1}{f} \qquad (4.93)$$

$$\xi_3' = \frac{\eta_1 \alpha_2}{f} \qquad (4.94)$$

The following considerations can be made:

1. Since $\eta_0 \ll f$, ξ_1' and ξ_2' can be reasonably approximated as follows:

$$\xi_1' \simeq \alpha_1 + \frac{\eta_1 \alpha_0}{f} \tag{4.95}$$

$$\xi_2' \simeq \alpha_2 + \frac{\eta_1 \alpha_1}{f} \tag{4.96}$$

2. It can be demonstrated that the second term in Equation (4.95) produces a shift along the Doppler coordinate. Therefore, Equation (4.95) can be approximated as follows:

$$\xi_1' \simeq \alpha_1 \tag{4.97}$$

3. When the following inequality is satisfied,

$$\left| \frac{4\pi}{c} \left(\eta_1 \alpha_1 T_{obs}^2 + \eta_1 \alpha_2 T_{obs}^3 \right) \right| < \frac{\pi}{4} \tag{4.98}$$

the focusing parameters ξ_2' in Equation (4.96) and ξ_3' in Equation (4.94) can be approximated as follows:

$$\xi_2' \simeq \alpha_2 \tag{4.99}$$

$$\xi_3' \simeq 0 \tag{4.100}$$

In order to ensure that the inequality in Equation (4.98) is satisfied, the following more restrictive constraints are applied:

$$|\eta_1| < \frac{c}{32 \, |\alpha_1| \, T_{obs}^2} \tag{4.101}$$

$$|\eta_1| < \frac{c}{32 \, |\alpha_2| \, T_{obs}^3} \tag{4.102}$$

Depending on the type of oscillators used, the frequency drift may vary from 1 part per million (10^{-6}) to 1 part per trillion (10^{-12}) or better within a temperature range of 50-100°. Low cost quartz oscillators with no temperature control have frequency drifts of 10^{-6} over a range of 50-100°, whereas temperature-controlled quartz oscillators can reduce the drift to 10^{-7} over the same temperature range. Oven-controlled quartz oscillators may improve to 10^{-8} and expensive rubidium or caesium oscillators may reach frequency drifts of 10^{-12} or better. Moreover, in typical ISAR applications, the coherent processing interval (CPI or T_{obs}) are quite long, of the order of 1 sec. Then, inequalities (4.101) and (4.102) are usually satisfied. The straightforward result is that, in practice, the parametric autofocusing techniques proposed in the literature for monostatic ISAR can be applied to B-ISAR in order to remove the bistatically equivalent radial motion, provided that the constraints in (4.101) and (4.102) are satisfied. Moreover, non-parametric autofocusing techniques, such as prominent point processing (PPP) [6] and phase gradient autofocus

[7, 5], as described in chapter 3, would not be affected at all by oscillator frequency offsets and drifts since they would treat the phase to be compensated as a random process. The consequent result is that non-parametric autofocusing techniques proposed for the monostatic case are applicable to the bistatic case.

4.5.3 B-ISAR IMAGE FORMATION IN THE PRESENCE OF SYNCHRONIZATION ERRORS

In this section the effects of synchronization errors on the B-ISAR image formation are analyzed. The received signal reflected by a single ideal scatterer located at a $\mathbf{x}_k$ can be written, after motion compensation and under far-field and short integration time condition, as in Equation (4.21).

Here, the synchronization errors effect is considered, then Equation (4.21) is rewritten as follows:

$$S_C(f,t) = W(f,n) \cdot \sigma_k \cdot e^{-j\varphi_x(x_{1_k},x_{2_k},f,t)} \tag{4.103}$$

where

$$\varphi_x(x_{1_k},x_{2_k},f,t) = \frac{4\pi}{c}\{[f + \eta_0 + \eta_1 t + \delta F(t)] \tag{4.104}$$
$$\cdot [K_0 + K_1 t] \cdot [x_{1_k}\sin(\Omega_B t) + x_{2_k}\cos(\Omega_B t)]\}$$

When the target undergoes small rotations and by neglecting the random phase term $\delta F(t)$, Equation (4.104) can be approximated by

$$\varphi_x(x_{1_k},x_{2_k},f,t) \simeq \frac{4\pi}{c}\{[f + \eta_0 + \eta_1 t] \cdot [K_0 + K_1 t] \cdot [x_{1_k}\Omega_B t + x_{2_k}]\} \tag{4.105}$$

4.5.3.1 Range Compression

The range compressed signal is obtained by calculating the IFT with respect to the variable f, as follows:

$$
S_{CR}(\tau,t) = FT_f^{-1}\{S_C(f,t)\} = \tag{4.106}
$$

$$
= \sigma_k \int W(f,t)\exp\left\{-j\frac{4\pi}{c}(f+\eta_0+\eta_1 t)(K_0+K_1 t)(x_{1_k}\Omega_B t+x_{2_k})\right\}e^{j2\pi f\tau}\mathrm{d}f =
$$

$$
= \sigma_k \exp\left\{-j\frac{4\pi(\eta_0+\eta_1 t)}{c}(K_0+K_1 t)(x_{1_k}\Omega_B t+x_{2_k})\right\}
$$

$$
\int_{-\infty}^{\infty} W(f,t)\exp\left\{-j2\pi f\left(\frac{2}{c}(K_0+K_1 t)(x_{1_k}\Omega_B t+x_{2_k})\right)\right\}e^{j2\pi f\tau}\mathrm{d}f =
$$

$$
= \sigma_k \exp\left\{-j\frac{4\pi(\eta_0+\eta_1 t)}{c}(K_0+K_1 t)(x_{1_k}\Omega_B t+x_{2_k})\right\}
$$

$$
\delta\left(\tau-\frac{2}{c}(K_0+K_1 t)(x_{1_k}\Omega_B t+x_{2_k})\right)\otimes_\tau \tilde{w}(\tau,t)=
$$

$$
= \sigma_k B\cdot\mathrm{rect}\left(\frac{t}{T_{obs}}\right)\mathrm{sinc}\left\{B\left(\tau-\frac{2}{c}(K_0 x_{2_k}+r_m(t))\right)\right\}e^{j2\pi f_0\tau}
$$

$$
\cdot\exp\left\{-j\frac{4\pi(f_0+\eta_0+\eta_1 t)}{c}(K_0+K_1 t)(x_{1_k}\Omega_B t+x_{2_k})\right\}
$$

where $\tilde{w}(\tau,t)$ is defined in Equation (4.23) and $r_m(t)$ in Equation (4.24). It must be pointed out that both the range compression and migration components are (1) dependent on bistatic angle and (2) independent on synchronization errors. Therefore, the same constraint as in Equation (4.25) should be applied to avoid range migration.

When the constraint in Equation (4.25) is satisfied, the range-compressed signal in Equation (4.106) can be approximated as follows:

$$
S_{CR}(\tau,t) \simeq \sigma_k\cdot B\cdot\mathrm{rect}\left(\frac{t}{T_{obs}}\right)\mathrm{sinc}\left\{B\left(\tau-\frac{2}{c}K_0 x_{2_k}\right)\right\} \tag{4.107}
$$

$$
e^{j2\pi f_0\tau}\cdot\exp\left\{-j\frac{4\pi(f_0+\eta_0+\eta_1 t)}{c}(K_0+K_1 t)(x_{1_k}\Omega_B t+x_{2_k})\right\}
$$

4.5.3.2 Cross-Range Compression

As shown in section 4.3.2, the cross-range compression is obtained by calculating the IFT with respect to the variable t (slow-time), as follows:

$$I(\tau, \nu, x_{1_k}, x_{2_k}) \quad = \quad FT_t^{-1}\left\{S_{CR}(\tau, t)\right\} = \tag{4.108}$$

$$= \sigma_k \cdot B \cdot \mathrm{sinc}\left\{B\left(\tau - \frac{2}{c}K_0 x_{2_k}\right)\right\} e^{j2\pi f_0 \tau}$$

$$\cdot \int_{-\infty}^{\infty} \mathrm{rect}\left(\frac{t}{T_{obs}}\right) \exp\left\{-j\frac{4\pi(f_0 + \eta_0 + \eta_1 t)}{c}(K_0 + K_1 t)(x_{1_k}\Omega_B t + x_{2_k})\right\} e^{j2\pi\nu t} dt$$

$$= \sigma_k \cdot B \cdot \mathrm{sinc}\left\{B\left(\tau - \frac{2}{c}K_0 x_{2_k}\right)\right\} e^{j2\pi f_0\left(\tau - \frac{2}{c}K_0 x_{2_k}\right)}$$

$$\cdot \int_{-\infty}^{\infty} \mathrm{rect}\left(\frac{t}{T_{obs}}\right) \exp\left\{-j\frac{4\pi(\eta_0 + \eta_1 t)}{c}(K_0 + K_1 t)(x_{1_k}\Omega_B t + x_{2_k})\right\}$$

$$\exp\left\{-j2\pi\left(\frac{2f_0}{c}\left((K_0 x_{1_k}\Omega_B + K_1 x_{2_k})t + K_1 x_{1_k}\Omega_B t^2\right)\right)\right\} e^{j2\pi\nu t} dt$$

$$= \delta\left(\tau - \frac{2}{c}K_0 x_{2_k}\right)\delta\left(\nu - \frac{2f_0\Omega_B}{c}K_0 x_{1_k}\right) \otimes_\tau \otimes_\nu w(\tau, \nu) \otimes_\nu D(\nu)$$

where $w(\tau, \nu)$ is defined in Equation (4.32) and

$$D(\nu) = \int_{-\infty}^{\infty} \exp\left\{-j\frac{4\pi}{c}\left(\gamma_0 + \gamma_1 t + \gamma_2 t^2 + \gamma_3 t^3\right)\right\} e^{j2\pi\nu t} dt \tag{4.109}$$

$$\gamma_0 \quad = \quad \eta_0 K_0 x_{2_k} \tag{4.110}$$

$$\gamma_1 \quad = \quad \eta_0 K_0 \Omega_B x_{1_k} + (\eta_1 K_0 + \eta_0 K_1) x_{2_k} + f_0 K_1 x_{2_k} \tag{4.111}$$

$$\gamma_2 \quad = \quad f_0 K_1 \Omega_B x_{1_k} + \eta_0 K_1 \Omega_B x_{1_k} + \eta_1\left(K_0 \Omega_B x_{1_k} + K_1 x_{2_k}\right) \tag{4.112}$$

$$\gamma_3 \quad = \quad \eta_1 K_1 \Omega_B x_{1_k} \tag{4.113}$$

summarizes the effects of cr_s defined in Equation (4.30), the chirp-like distortion term defined in Equation (4.31) and the distortions caused by the synchronization errors (η_0 and η_1). In particular, the term $f_0 K_1 x_{2_k}$ in Equation (4.111) corresponds to the cr_s defined in Equation (4.30), whereas $f_0 K_1 \Omega_B x_{1_k}$ in Equation (4.112) is the chirp-like distortion term defined in Equation (4.31). Therefore, the distortion term in Equation (4.109) can be written in terms of a linear phase term and a quadratic/cubic phase term as follows:

$$D(\nu) = D_1(\nu) \otimes_\nu D_2(\nu) \cdot e^{j\theta_0} \tag{4.114}$$

where

$$D_1(\nu) = \int_{-\infty}^{\infty} \exp\left\{-j\frac{4\pi}{c}\gamma_1 t\right\} e^{j2\pi\nu t} dt = \delta\left(\nu - \frac{2}{c}\gamma_1\right) \quad (4.115)$$

$$D_2(\nu) = \int_{-\infty}^{\infty} \exp\left\{-j\frac{4\pi}{c}\left(\gamma_2 t^2 + \gamma_3 t^3\right)\right\} e^{j2\pi\nu t} dt \quad (4.116)$$

$$\theta_0 = -\frac{4\pi}{c}\eta_0 K_0 x_{2_k} \quad (4.117)$$

$D_1(\nu)$ in Equation (4.115) is the linear phase term, whereas $D_2(\nu)$ in Equation (4.116) is the quadratic/cubic one. The term γ_0 does not produce any distortions or shift since the phase term, $-4\pi\gamma_0/c$, is constant.

4.5.3.3 Comments on the Bistatic PSF in the Presence of Synchronization Errors

It should be noted that the position of the scattering center in the Doppler-delay time domain, represented by the two delta functions in Equation (4.108), is scaled with respect to the monostatic case by the bistatic factor K_0. Some more remarks follow:

1. The PSF of the range-Doppler B-ISAR image, in the presence of synchronization errors, can be interpreted as a distorted version of the monostatic PSF.
2. A linear space-variant distortion term, $D_1(\nu)$, causes scattering center shifts along the Doppler coordinate that depend on their spatial position (both range and cross-range coordinates). Although this term looks like a shift term, it leads to an image distortion because it introduces a spatial-dependent scatterer's displacement.
3. A non-linear space-variant distortion term, $D_2(\nu)$, causes chirp-like and third-order distortions that result in PSF smearing (typical defocusing effect).

4.5.4 DISTORTION ANALYSIS

A detailed distortion analysis that aims at quantifying the effects of $D_1(\nu)$ and $D_2(\nu)$ follows. This analysis provides a quantitative analysis of the distortions introduced by the joint action of bistatic angle changes and phase synchronization errors.

4.5.4.1 Linear Distortions Analysis

The linear distortion term can be quantified in terms of relative Doppler shift and can be compared with the Doppler resolution for assessing its impact on the B-ISAR image. Specifically, the Doppler-induced shift caused by the linear distortion term can be expressed as in Equation (4.118).

$$\Delta_1 = \frac{2}{c} \{\eta_0 K_0 \Omega_B x_{1_k} + [\eta_1 K_0 + (f_0 + \eta_0) K_1] x_{2_k}\} \qquad (4.118)$$

$$\simeq \frac{2}{c} \{\eta_0 K_0 \Omega_B x_{1_k} + [\eta_1 K_0 + f_0 K_1] x_{2_k}\}$$

where the approximation is accurate since $\eta_0 \ll f_0$.

As pointed out above, the Doppler shift (cross-range shift) depends on both the scatterer spatial coordinates in the projected image plane. Nevertheless, it can be argued that the linear distortion in Equation (4.118) can be neglected when its absolute value is smaller than that of the Doppler resolution cell:

$$|\Delta_1| < \frac{1}{T_{obs}} \qquad (4.119)$$

By applying Equation (4.119) and after simple manipulations, Equation (4.118) can be written as follows:

$$|\eta_0 K_0 \Omega_B x_{1_k} + [\eta_1 K_0 + f_0 K_1] x_{2_k}| <$$
$$|\eta_0 K_0 \Omega_B x_{1_k}| + |\eta_1 K_0 x_{2_k}| + |f_0 K_1 x_{2_k}| < \frac{c}{2 T_{obs}} \qquad (4.120)$$

It can be noted that, in the worst bistatic scenario ($K_0 = 1$, which represents the monostatic case), and even considering poor performance oscillators, the following two components can be disregarded:

$$|\eta_0 x_{1_k}| < \frac{c}{2 \Omega_B T_{obs}} \qquad (4.121)$$

$$|\eta_1 x_{2_k}| < \frac{c}{2 T_{obs}} \qquad (4.122)$$

Therefore, the remaining constraint can be expressed as follows:

$$|K_1 x_{2_k}| < \frac{c}{2 f_0 T_{obs}} \qquad (4.123)$$

which is purely a geometrical constraint. In other words, synchronization errors, even when coupled with bistatic angle changes, do not cause significant linear distortions. It is also important to note that the Doppler displacement caused by the dominant component Equation (4.123) is linearly dependent on the range coordinate (x_{2_k}) and on the bistatic angle change rate (K_1).

$$\Delta_1 \simeq \frac{2}{c} f_0 K_1 x_{2_k} \qquad (4.124)$$

By comparing Equation (4.124) with Equation (4.30) and Equation (4.74) it can be noted that, given a bistatic angle change rate, the Doppler displacement becomes proportional to the range coordinate. This is an important result as the linear distortion given by the term in Equation (4.124) only causes a linear range-dependent Doppler shift. The visible effect for small values of the Doppler shift in Equation (4.124) is an apparent target rotation in the B-ISAR image domain, although it is not a physical rotation. Given that the linear distortion only produces such an effect, the condition in Equation (4.123) can be relaxed.

4.5.4.2 Quadratic Distortions Analysis

The quadratic term causes chirp-like distortions in the B-ISAR image. To limit the distortions to an acceptable level, the following constraint must be taken into account:

$$|\Delta_2| < \frac{1}{T_{obs}^2} \tag{4.125}$$

where

$$\Delta_2 \doteq \frac{2}{c}\gamma_2 = \frac{2}{c}\left[(f_0 + \eta_0)\,K_1\Omega_B x_{1_k} + \eta_1\,(K_0\Omega_B x_{1_k} + K_1 x_{2_k})\right] \tag{4.126}$$

By approximating $(f_0 + \eta_0) \simeq f_0$ and by considering the worst bistatic scenario $(K_0 = 1)$, the following more restrictive constraint can be formulated:

$$\left|\frac{2}{c}f_0 K_1\Omega_B x_{1_k}\right| + \left|\frac{2}{c}\eta_1\Omega_B x_{1_k}\right| + \left|\frac{2}{c}\eta_1 K_1 x_{2_k}\right| < \frac{1}{T_{obs}^2} \tag{4.127}$$

which can be split into three separate constraints by constraining each term to be less than a third of $1/T_{obs}^2$, as follows:

$$|K_1 x_{1_k}| \quad < \quad \frac{c}{6 f_0 \Omega_B T_{obs}^2} \tag{4.128}$$

$$|\eta_1 x_{1_k}| \quad < \quad \frac{c}{6 \Omega_B T_{obs}^2} \tag{4.129}$$

$$|\eta_1 K_1 x_{2_k}| \quad < \quad \frac{c}{6 T_{obs}^2} \tag{4.130}$$

The following must be pointed out.

1. The constraint in Equation (4.128) is a geometrical-only constraint, since it only depends on the bistatic angle (compare with Equation (4.33));
2. The constraint in Equation (4.129) only depends on synchronization errors;
3. The constraint in Equation (4.130) depends on both the bistatic angle and synchronization errors.

It is important to note that the inequalities in Equations (4.128)-(4.130) can be neglected unless very rapid bistatic angle changes and very large synchronization errors occur.

4.5.4.3 Cubic Distortions Analysis

In order to limit cubic distortions, the following constraint applies:

$$|\Delta_3| < \frac{1}{T_{obs}^3} \tag{4.131}$$

where

$$\Delta_3 \doteq \frac{2}{c}\gamma_3 = \frac{2}{c}\eta_1 K_1 \Omega_B x_{1_k} \tag{4.132}$$

Also the quantity $|\Delta_3|$ can be neglected unless very rapid bistatic angle changes occur in conjunction with very large synchronization errors. The inequality in Equation (4.131) is rewritten in Equation (4.133) in order to highlight the joint dependence on synchronization errors and bistatic angle changes.

$$|\eta_1 K_1 x_{1_k}| < \frac{c}{2\Omega_B T_{obs}^3} \tag{4.133}$$

Although the joint action of bistatic geometry changes and phase synchronization errors cause distortions, for these to be noticeable in the B-ISAR image, particular scenarios with strong bistatic angle changes and large synchronization errors must be considered.

In the rare event of significant quadratic distortions, the quadratic distortion term can be nullified by using time-frequency distributions (TFD) instead of the FT to form the image along the Doppler coordinate, as has been formally demonstrated in a number of papers [16, 4, 3].

A few examples that aim at verifying the effectiveness of monostatic ISAR autofocusing techniques when used for bistatic ISAR and analyzing the distortion effects caused by both bistatic angle changes and synchronization errors when using a monostatic ISAR processor can be found in [11].

4.6 CONCLUSIONS

In this chapter the bistatic ISAR problem has been investigated and the received signal model defined. A bistatic ISAR image reconstruction algorithm has been defined which makes use of the monostatic ISAR processing and the definition of the bistatically equivalent monostatic (BEM) radar. Moreover, the main differences between the monostatic and bistatic cases have been investigated, among them the distortions introduced by the bistatic geometry. More specifically, two sources of problems affecting the bistatic ISAR images, namely bistatic angle changes and synchronization errors, have been deeply investigated. The main results of this chapter are

- The applicability of the monostatic ISAR processor to the bistatic geometry in most cases.
- A methodology to assess the applicability of monostatic ISAR processor to arbitrary bistatic ISAR scenario.

- An analysis of the distortions caused by the bistatic geometry and synchronization errors allowing for the correct interpretation of bistatic ISAR images.

REFERENCES

1. F. Berizzi and E. Dalle Mese. Sea-wave fractal spectrum for sar remote sensing. *Radar, Sonar and Navigation, IEE Proceedings -*, 148(2):56–66, 2001.
2. F. Berizzi, M. Martorella, B. Haywood, E. Dalle Mese, and S. Bruscoli. A survey on isar autofocusing techniques. *Image Processing, 2004. ICIP '04. 2004 International Conference on*, 1:9–12, 2004.
3. F. Berizzi, E. D. Mese, M. Diani, and M. Martorella. High-resolution isar imaging of maneuvering targets by means of the range instantaneous doppler technique: modeling and performance analysis. *Image Processing, IEEE Transactions on*, 10(12):1880–1890, 2001.
4. V. C. Chen and H. Ling. *Time-Frequency Transforms for Radar Imaging and Signal Analysis*. Artech House Mobile Communications Series. Artech House, 1st edition, January 2002.
5. P. H. Eichel and C. V. Jakowatz. Phase-gradient algorithm as an optimal estimator of the phase derivative. *Optics Letters*, 14(20):1101–1103, October 1989.
6. B. Haywood and R. J. Evans. Motion compensation for isar imaging. *In Proceedings of the Australian Symposium on Signal Processing and Applications (ASSPA 89), Adelaide, Australia*, pages 113–117, 1989.
7. C. V. J. Jakowatz, D. E. Wahl, P. H. Eichel, D. C. Ghiglia, and P. A. Thompson. *Spotlight-Mode Synthetic Aperture Radar: A Signal Processing Approach*. New York: Springer (Ed.), 1996.
8. G. Krieger and M. Younis. Impact of oscillator noise in bistatic and multistatic sar. *Geoscience and Remote Sensing Letters, IEEE*, 3(3):424–428, 2006.
9. P. Lopez-Dekker, J.J. Mallorqui, P. Serra-Morales, and J. Sanz-Marcos. Phase synchronization and doppler centroid estimation in fixed receiver bistatic sar systems. *Geoscience and Remote Sensing, IEEE Transactions on*, 46(11):3459–3471, 2008.
10. M. Martorella. Optimal sensor positioning for isar imaging. *Geoscience and Remote Sensing Symposium (IGARSS), 2010 IEEE International*, pages 4819–4822, 2010.
11. M. Martorella. Analysis of the robustness of bistatic inverse synthetic aperture radar in the presence of phase synchronisation errors. *Aerospace and Electronic Systems, IEEE Transactions on*, 47(4):2673–2689, 2011.
12. M. Martorella, F. Berizzi, and B. Haywood. Contrast maximisation based technique for 2-d isar autofocusing. *Radar, Sonar and Navigation, IEE Proceedings -*, 152(4):253–262, 2005.
13. M. Martorella, D. Cataldo, and S. Brisken. Bistatically equivalent monostatic approximation for bistatic isar. *Radar Conference (RADAR), 2013 IEEE*, pages 1–5, 2013.
14. M. Martorella, J. Palmer, F. Berizzi, and B. Bates. Advances in bistatic inverse synthetic aperture radar. *Radar Conference - Surveillance for a Safer World, 2009. RADAR. International*, pages 1–6, 2009.

15. M. Martorella, J. Palmer, J. Homer, B. Littleton, and I. D. Longstaff. On bistatic inverse synthetic aperture radar. *Aerospace and Electronic Systems, IEEE Transactions on*, 43(3):1125–1134, 2013.

16. T. Thayaparan, L. Stankovic, C. Wernik, and M. Dakovic. Real-time motion compensation, image formation and image enhancement of moving targets in isar and sar using s-method based approach. *Signal Processing, IET*, 2(3):247–264, 2008.

17. W. Q. Wang. *Bistatic Synthetic Aperture Radar Synchronization Processing*. Radar Technology Series. Guy Kouemou (Ed.), January 2010. Available online at

http://www.intechopen.com/books/radar-technology/
bistatic-synthetic-aperture-radar-synchronization-processing.

18. W. Q. Wang and D. Cai. Antenna directing synchronization for bistatic synthetic aperture radar systems. *Antennas and Wireless Propagation Letters, IEEE*, 9:307–310, 2010.

5 Passive ISAR

*M. Martorella, F. Berizzi, E. Giusti, C.
Moscardini, A. Capria, D. Petri, and M. Conti*

CONTENTS

Passive radars (PRs) have drawn the attention of the scientific community
for many decades, as they offer a number of advantages over conventional
active radar systems [14, 9, 6, 10, 8, 16, 2, 7]. A passive radar is intrinsically
a bistatic radar, since the transmitter and the receiver are not co-located.
Differently from the bistatic radar, PRs exploit illuminators of opportunity
(IOs) as electromagnetic sources to illuminate targets of interest. The use
of non-cooperative IOs imposes the need to use two receiving channels so
as to perform the matched filter at the receiver. One channel gathers the
reference signal, namely the transmitted signal, so as to have a copy of the
transmitted signal at the receiver, while the other channel gathers the echoes
from all the targets in the surveillance area, namely the received signal. A
pictorial representation of how a PR works is depicted in Figure 5.1, where
at the receiver location two antennas define both the reference channel and
the surveillance channel. The bistatic geometry offers an enhanced resilience
against electronic counter measure (ECM) with respect to active radars and

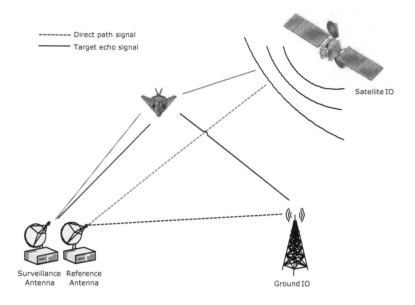

Figure 5.1 Passive radar principle with two different IOs, a ground-based broadcast station and a spaceborne IO

a solution against stealth technology, which is designed primarily to defeat monostatic radars.

Moreover, dealing with passive radar systems, the use of IOs reduces the overall cost of the system. Typically IOs are classified as either external dedicated radar systems or other types of illuminators, such as broadcast transmitters. The latter are of great interest as they allow large areas to be covered, a wide range of frequencies to be used, which are not normally available for active radar systems, and signals to be acquired continuously.

Nonetheless, some drawbacks are also worth mentioning, and especially that the location of the transmitter and the waveform are not under the control of the radar designer.

As research in this field progresses, more radar techniques are added to PRs to make them able to handle several tasks and to be applied in different scenarios. One such task is the radar imaging of non-cooperative targets through inverse synthetic aperture radar (ISAR) imaging, which in turn may open the doors to non-cooperative target recognition (NCTR) capabilities [17, 11].

The purpose of this chapter is to pave the theoretical foundation of the theory behind the passive ISAR imaging technique. However, to reach this goal, the passive radar theory must be introduced first as well as the mathematical formulation of the passive radar ambiguity function.

The reminder of the chapter is organized as follows. Section 5.1 introduces the concept of PR. Section 5.2 deals with the range Doppler map formation process. Section 5.3 details the passive ISAR algorithm whereas Section 5.4 provides an analysis of different IOs in terms of ISAR image performances. Conclusions are finally drawn in Section 5.5.

5.1 PASSIVE RADAR SIGNAL PROCESSING CHAIN

PR systems exploit reflection from IOs in order to detect and track objects. The PR performance in terms of both target detection and imaging are strictly dependent on the IO properties, and specifically on the emitted power density, which impact on the receiver range coverage and on the SNR at the receiver, and the transmitted waveform characteristics.

As already said in the introduction, a problem of PR systems with respect to monostatic radar systems, concerns the fact that there is no copy of the transmitted signal at the receiver. This problem is important for every bistatic configuration, but it becomes relevant for passive configurations because the transmitter is non-cooperative. Therefore, a PR receiver requires at least two channels to perform the matched filtering operation: the reference channel which receives a copy of the transmitted signal at the receiver, and the surveillance channel which receives echoes from targets within the surveillance area. Therefore, the most simple PR system is composed of two highly directive dedicated antennas used for the acquisition of the surveillance and the reference signals, as shown in Figure 5.1. Matched filtering is achieved by correlating the surveillance signal $s_R(t)$ with Doppler-shifted versions of the reference signal $s_{ref}(t)$ to form a range-Doppler map. The range-Doppler map is also called the cross ambiguity function (CAF), namely the cross-correlation between the reference and the surveillance signals. The ambiguity function (AF) is instead the auto-correlation function of the reference signal. The matched filter stage is needed for two important purposes: obtaining a sufficient signal processing gain to allow the targets to be detected above the noise floor and estimating the bistatic range and bistatic Doppler shift of the target echoes. This solution based on the matched filter is the optimum receiver in the presence of additive white noise, but a typical PR scenario is more complex and this simple structure presents several drawbacks. An advanced signal processing architecture is usually adopted in order to mitigate and ideally suppress the several interference signals received on the surveillance channel to a level under the noise floor. A general block scheme of a PR processing chain is sketched in Figure 5.2 and it consists of the following processing steps:

1. Reference signal pre-processing block. This block aims at improving the quality of the AF and then of the CAF. More specifically this block is devoted to removing spurious peaks in the ambiguity function caused by deterministic and pseudo-random pilot tones and

reduce the multipath effects in the reference signal. Pilot tones may be present in both analog (analog TV, for example) and digital wave-forms (DVB-T, DVB-S, DVB-SH, etc.) and are typically used for synchronization purposes. For what concerns the PR system and its ambiguity function, such pilot tones determine spurious peaks which if present in the RD map may affect the detection performance. Multipath is instead due to the fact that the reference signal acquired by means of the reference channel does not necessarily come from the main path, namely the direct path, but conversely, it may be the superposition of two or more paths generated by reflection from ground or other structures. Multipath in the reference signal may affect the detection performance as shown in [22]. An effective way to mitigate multipath effects consist in demodulating the reference signal, re-covering the transmitted QAM (quadrature amplitude modulation) symbols and then remodulating such symbols so as to reproduce a reference signal free of multipath.

2. Interference suppression. This block is used to filter out the direct path interference, its multipath and the ground clutter, which are all characterized by a zero Doppler spectrum in the range-Doppler domain. A wide variety of temporal adaptive processing techniques have been developed for the removal of the interferences such as that in [19].

3. Matched filter. It is used to generate the range Doppler maps to determine target bistatic range and Doppler. This block represents the key processing step in a passive radar [23].

4. Detector and tracker. Such block is devoted to detect and track tar-gets in the range-Doppler domain. The target tracking in spatial co-ordinates can be defined by exploiting measurements of the bistatic range, the bistatic velocity and the direction of arrival of a target [20].

5.2   RANGE DOPPLER MAP FORMATION

In a passive radar system, the optimum way to process the received signal is given by the theoretical CAF calculation [27], defined as

$$\chi(\tau,\nu) = \int_0^{T_{int}} s_R(t) s_{ref}{}^*(t-\tau) e^{-j2\pi\nu t} dt \tag{5.1}$$

where $\chi(\tau,\nu)$ represents the range-Doppler cross-correlation function between the reference signal $s_{ref}(t)$ and the surveillance signal $s_R(t)$, the variable τ denotes the delay-time, ν corresponds to the frequency Doppler shift, T_{int} de-notes the integration time or the so-called coherent processing interval (CPI). In order to avoid integration losses, the integration time is typically chosen

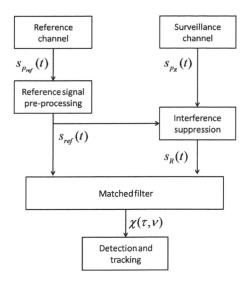

Figure 5.2 Signal processing chain for PBR

equal to $T_{int} = T_{obs} + \tau_{\max}$, where T_{obs} is the time interval for which the reference signal has been acquired and τ_{max} is the maximum delay time which corresponds to the maximum non-ambiguous bistatic range. The most obvious way to carry out the cross-correlation is to calculate the Fourier transform of $s_R(t)s_{ref}{}^*(t - \tau)$, known in literature as the mixing product, for each bistatic time delay. Another way is to calculate the cross-correlation between $s_R(t)$ and $s_{ref}(t - \tau)$ for each Doppler shift. These two approaches are known in literature as the Fourier transform based method and cross-correlation based method, respectively [3]. The computation of the CAF (or the RD map) through Equation (5.1) is however very computational. To save computational efforts, sub-optimum implementations of Equation (5.1) so as to achieve real time processing capabilities have been proposed in the past by several authors [26, 9, 18, 3]. Among all these sub-optimum approaches, the one proposed in [24, 15] deserves to be mentioned because it represents an excellent trade-off between computational effort and sub-optimality. Such algorithm focuses on the so-called *like-FMCW algorithm* or *batch algorithm*. Specifically, in [24, 15] a new revised formulation of the optimum CAF has been proposed followed by a new generalized and detailed mathematical formulation of the sub-optimum *batch algorithm*.

5.2.1   OPTIMUM CAF CALCULATION VIA THE BATCH ALGORITHM

The reference signal $s_{ref}(t)$, in Equation (5.1), can be seen as the sum of n_B contiguous batches of length T_B and it can be written as

$$s_{ref}(t) = \sum_{n=0}^{n_B-1} s_{refb}(t - nT_B, n) \tag{5.2}$$

where n_B is the number of batches obtained as

$$n_B = \left\lfloor \frac{T_{obs}}{T_B} \right\rfloor \tag{5.3}$$

where $\lfloor . \rfloor$ indicates the maximum integer value that is smaller than the argument. It should be noted that T_{obs} is the temporal length of the reference signal. The signal in each block is defined as

$$s_{refb}(t, n) = s_{ref}(t + nT_B) q(t) \tag{5.4}$$

where $q(t)$ is

$$q(t) = \begin{cases} 1 & t \in [0, T_B] \\ 0 & otherwise \end{cases} \tag{5.5}$$

By substituting Equation (5.2) into Equation (5.1), the optimum CAF can be written as

$$\chi(\tau, \nu) = \sum_{n=0}^{n_B-1} \int_0^{T_{int}} s_R(t) s_{refb}^*(t - \tau - nT_B, n) e^{-j2\pi\nu t} dt \tag{5.6}$$

According to the definition of $s_{refb}(t, n)$ in Equation (5.4), the integral in Equation (5.6) can be modified as follows:

$$\chi(\tau, \nu) = \sum_{n=0}^{n_B-1} \int_{nT_B+\tau}^{nT_B+\tau+T_B} s_R(t) s_{refb}^*(t - \tau - nT_B, n) e^{-j2\pi\nu t} dt \tag{5.7}$$

considering that $s_{refb}^*(t - \tau - nT_B, n) \neq 0$ $\forall t \in [nT_B + \tau, nT_B + \tau + T_B]$.

Setting $\alpha = t - nT_B$ and considering that the target distance is not a priori known, $\tau \in [0, \tau_{\max}]$, Equation (5.6) can be reformulated as

$$\chi(\tau, \nu) = \sum_{n=0}^{n_B-1} e^{-j2\pi\nu nT_B} \int_0^{T_B+\tau_{\max}} s_{Rb}(\alpha, n) s_{refb}^*(\alpha - \tau, n) e^{-j2\pi\nu\alpha} d\alpha \tag{5.8}$$

where $s_{Rb}(t, n)$ is the nth block of the surveillance signal $s_R(t)$, which is defined as

$$s_{Rb}(t, n) = s_R(t + nT_B) r(t) \tag{5.9}$$

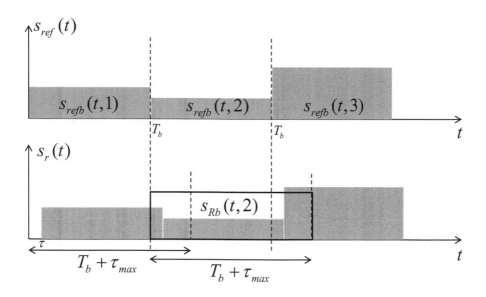

Figure 5.3 Graphical representation of reference and surveillance signal segmentation

where $r(t)$ is

$$r(t) = \begin{cases} 1 & t \in [0, T_B + \tau_{\max}] \\ 0 & otherwise \end{cases} \tag{5.10}$$

In order to guarantee the maximum integration gain at the output of each cross-correlation, the signal $s_R(t)$ is divided into partially overlapped batches. For the sake of clarity, the way in which both $s_{ref}(t)$ and $s_R(t)$ are split into batches of shorter lengths is graphically represented in Figure 5.3.

By considering the signals $s_{Rb}(t, n)$ and $s_{refb}(t, n)$, the CAF relative to the signals in the nth batch, $\chi_n(\tau, \nu)$, can be defined as

$$\chi_n(\tau, \nu) = \int_{-\infty}^{+\infty} s_{Rb}(t, n) \, s_{refb}^*(t - \tau, n) \, e^{-j2\pi\nu t} dt \tag{5.11}$$

Then, by substituting Equation (5.11) into Equation (5.8) yields

$$\chi_{opt}(\tau, \nu) = \sum_{n=0}^{n_B - 1} e^{-j2\pi\nu n T_B} \chi_n(\tau, \nu)$$
$$0 \leq \tau \leq \tau_{\max}$$
$$-\nu_{\max} \leq \nu \leq \nu_{\max} \tag{5.12}$$

Therefore, the optimum cross-ambiguity function $\chi_{opt}(\tau, \nu)$, between $s_R(t)$ and $s_{ref}(t)$, can be seen as a weighed sum of the cross-ambiguity functions,

$\chi_n(\tau, \nu)$, evaluated within each batch. Up to this stage, only a new formulation of the optimum CAF has been suggested without gaining with respect to the computational load and time processing.

5.2.2 BATCH ALGORITHM IMPLEMENTATION

A mathematical definition of the sub-optimum batches algorithm is derived by considering a specific approximation in the Doppler domain. Specifically, in Equation (5.11), when the product between T_B and the maximum target Doppler $\nu_{\max}$ is small compared to the unity, the phase rotation can be approximated as a constant value within each block. Specifically, the value of such a phase rotation is closer to the central time of the batch interval $(T_B/2)$

$$2\pi\nu t \approx 2\pi\nu\frac{T_B}{2} \quad \forall t \in [0, T_B], \nu \in [-\nu_{\max}, \nu_{\max}] \tag{5.13}$$

and Equation (5.12) can be simplified as

$$\chi_b(\tau, \nu) = e^{-j\pi\nu T_B} \sum_{n=0}^{n_B-1} e^{-j2\pi\nu n T_B} \int_{0}^{T_B+\tau_{\max}} s_{Rb}(\alpha, n) s_{refb}^*(\alpha - \tau, n) d\alpha \tag{5.14}$$

It should be noted that the target Doppler shift is estimated based on the increasing phase shift between consecutive batches. Due to the sampling theorem, this yields the relation $\nu_{\max} \leq 1/2T_B$. By defining the cross-correlation within the nth block as

$$\chi_{cc}(\tau, n) = \int_{0}^{T_B+\tau_{\max}} s_{Rb}(\alpha, n) s_{refb}^*(\alpha - \tau, n) d\alpha \tag{5.15}$$

Equation (5.14) can be written as

$$\chi_b(\tau, \nu) = e^{-j2\pi\nu T_B} \sum_{n=0}^{n_B-1} e^{-j2\pi\nu n T_B} \chi_{cc}(\tau, n) \tag{5.16}$$

or alternatively

$$\chi_b(\tau, \nu) = e^{-j2\pi\nu T_B} \sum_{n=0}^{n_B-1} e^{-j2\pi\nu n T_B} \chi_n(\tau, 0) \tag{5.17}$$

The Doppler frequency compensation is neglected inside each block while it is considered with respect to consecutive blocks. The cross-correlation term, $\chi_{cc}(\tau, n)$, in Equation (5.16) implements the matched filter or the pulse compression. It is therefore straightforward to conclude that it represents the range profiles relative to the nth block.

It is important to point out how the batch algorithm presents analogies with the classical pulse-Doppler processing used in active coherent radars. Pulse-Doppler processing provides a significant reduction in computational cost while suffering only a modest loss in processing gain. The received signal is first pulse-compressed by correlating the fast-time signals with the zero-Doppler matched filter. The range-Doppler map is then formed by taking the Fourier transform along the slow time domain. This processing is optimum under the assumption that the received waveform is an amplitude-scaled and time-delayed version of the transmitted waveform. The Doppler shift of the received waveform represents an unintentional mismatch between the received signal and the matched filter. The response of a waveform in the presence of an uncompensated Doppler is known in literature as Doppler tolerance. In an active radar system, the Doppler shift and its impact on the matched filter output signal may be approached in several ways. The first way is to select a waveform and its modulation parameters in order to achieve a Doppler-tolerant waveform. The main difference is that, in the active radar case, the transmitted waveform properties can be selected in order to obtain a Doppler tolerant waveform, while in a passive radar scenario only an appropriate batch length can be selected to reduce these losses.

5.2.3   PERFORMANCE ANALYSIS

The losses of the batch algorithm with respect to the optimum algorithm are evaluated by means of the improvement factor (IF) defined as follows:

$$IF(\tau, \nu) = 20\log_{10}\left(\left|\frac{SNR_b(\tau, \nu)}{SNR_{opt}(\tau, \nu)}\right|\right) \tag{5.18}$$

where $SNR_b(\tau, \nu)$ and $SNR_{opt}(\tau, \nu)$ represent the SNR obtained, respectively, by using the batch algorithm and the optimum one, respectively. Let the SNR be the ratio of the signal power to the mean noise power:

$$SNR(\tau, \nu) = \frac{\left|\chi^{(s)}(\tau, \nu)\right|^2}{E\left\{\left|\chi^{(w)}(\tau, \nu)\right|^2\right\}} \tag{5.19}$$

where $\chi^{(s)}(\tau, \nu)$ and $\chi^{(w)}(\tau, \nu)$ are the target and noise components obtained at the output of the CAF processing, respectively, and $E\{.\}$ represents the statistical expectation operator.

Let $\chi_b^{(s)}(\tau, \nu)$ and $\chi_{opt}^{(s)}(\tau, \nu)$ be the target components of RD map obtained at the output of the batch algorithm and the optimum algorithm, respectively.

In order to derive a closed form solution of $IF(\tau, \nu)$, both $SNR_b(\tau, \nu)$ and $SNR_{opt}(\tau, \nu)$ should be computed first.

For the sake of simplicity, let us assume the target composed of a single slowly fluctuating point scatterer. The received echo of such a target on the surveillance channel, after demodulation and baseband conversion, can be approximated as follows:

$$s_R(t) = \sigma_k \cdot s_{ref}(t - \tau_k)e^{j2\pi\nu_k t} \tag{5.20}$$

where τ_k and ν_k are the delay time and Doppler shift of the single point target and σ_k is a complex amplitude which is a function of the target reflectivity, propagation effects and antenna gain.

By segmenting the surveillance signal as done in Equation (5.2) and as graphically represented in Figure 5.3, the surveillance received signal in a batch can be expressed as follows:

$$
\begin{aligned}
s_{Rb}(t, n) = {} & \sigma_k \cdot e^{j2\pi\nu_k(t+nT_B)} \sum_{m_1=1}^{n-1} s_{refb}(t - \tau_k + m_1 T_B, n - m_1) + \\
& \sigma_k \cdot e^{j2\pi\nu_k(t+nT_B)} s_{ref}(t - \tau_k, n) + \\
& \sigma_k \cdot e^{j2\pi\nu_k(t+nT_B)} \sum_{m_2=n+1}^{n_B} s_{refb}(t - \tau_k - m_2 T_B, n + m_2)
\end{aligned} \tag{5.21}
$$

As can be noted, $s_{Rb}(t, n)$ is composed of two main components. The one is $s_{refb(t-\tau_k,n)}e^{j2\pi\nu_k(t+nT_B)}$ which is the desired signal, namely, the reference signal in the nth batch delayed by the target delay-time. The other term, namely

$$
\begin{aligned}
s_{Rb}^{CT}(t, n) = {} & \sigma_k e^{j2\pi\nu_k(t+nT_B)} \sum_{m_1=1}^{n-1} s_{refb}(t - \tau_k + m_1 T_B, n - m_1) + \\
& \sigma_k e^{j2\pi\nu_k(t+nT_B)} \sum_{m_2=n+1}^{n_B} s_{refb}(t - \tau_k - m_2 T_B, n + m_2)
\end{aligned} \tag{5.22}
$$

is instead due to the reference signal in the adjacent batches.

By substituting $s_{Rb}(t, n)$ in Equation (5.11), the signal component of $\chi_n(\tau, \nu)$ can be derived as follows:

$$
\begin{aligned}
\chi_n^{(s)}(\tau, \nu) &= e^{j2\pi\nu_T(t+nT_B)} \int_{-\infty}^{\infty} s_{Rb}(\alpha, n)s_{refb}^*(\alpha - \tau, n)e^{-j2\pi\nu t}dt \\
&= M_n(\tau, \nu) + M_n^{CT}(\tau, \nu)
\end{aligned} \tag{5.23}
$$

where

$$M_n(\tau, \nu) = \sigma_k e^{j2\pi\nu_k nT_B} \int_{-\infty}^{\infty} s_{refb}(t - \tau_k, n) \cdot s_{refb}^*(t - \tau, n)e^{-j2\pi(\nu-\nu_k)t}dt \tag{5.24}$$

and

$$M_n^{CT}(\tau,\nu) = \int_{-\infty}^{\infty} s_{Rb}^{CT}(t,n) \cdot s_{refb}^*(t-\tau,n)e^{-j2\pi\nu t}dt \qquad (5.25)$$

The contribution of $M_n(\tau,\nu)$ is larger than $M_n^{CT}(\tau,\nu)$ since the signals in different batches are typically uncorrelated. In any case, even when a small correlation exists, $M_n^{CT}(\tau,\nu)$ can be neglected as its average is much smaller than the average of $M_n(\tau,\nu)$.

Therefore, $M_n^{CT}(\tau,\nu)$ can be neglected as shown in Equation (5.26):

$$\chi_n^{(s)}(\tau,\nu) = \sigma_k e^{j2\pi\nu_k nT_B} \int_{-\infty}^{\infty} s_{refb}(t-\tau_k,n)s_{refb}^*(t-\tau,n)e^{-j2\pi(\nu-\nu_k)t} \quad (5.26)$$

by setting $\alpha = t - \tau_k$, the following equations can be derived:

$$\chi_n^{(s)}(\tau,\nu) = \sigma e^{j2\pi\nu_k nT_B} \int_{-\infty}^{\infty} s_{refb}(\alpha,n)s_{refb}^*(\alpha+\tau_k-\tau,n)e^{-j2\pi(\nu-\nu_k)(\alpha+\tau_k)} \, d\alpha$$
$$(5.27)$$

Then, simple mathematical manipulation leads to the following equation:

$$\chi_n^{(s)}(\tau,\nu) = \sigma_k e^{-j2\pi\nu_k nT_B} e^{j2\pi(\nu-\nu_k)\tau_k} A_n(\tau-\tau_k,\nu-\nu_k) \qquad (5.28)$$

where

$$A_n(\tau,\nu) = \int_{-\infty}^{\infty} s_{refb}(\alpha,n)s_{refb}^*(\alpha-\tau,n)e^{-j2\pi\nu\alpha} \, d\alpha \qquad (5.29)$$

Finally, by using Equation (5.12), the signal component of the optimum cross-ambiguity function can be derived as follows:

$$\chi_{opt}^{(s)} = \sigma_k e^{-j2\pi(\nu-\nu_k)\tau_k} \sum_{n=0}^{n_B-1} e^{-j2\pi(\nu-\nu_k)nT_B} A_n(\tau-\tau_k,\nu-\nu_k) \qquad (5.30)$$

Similar steps can be performed to obtain the signal component of cross-ambiguity function at the output of the batch algorithm. Specifically, starting from Equation (5.11) and by neglecting the cross terms, namely, $M_n^{CT}(\tau,\nu)$, $\chi_n^s(\tau,0)$ can be derived as follows:

$$\chi_n^{(s)}(\tau,0) = \sigma_k e^{j2\pi\nu_k nT_B} \int_{-\infty}^{\infty} s_{refb}(t-\tau_k,n)s_{refb}^*(t-\tau,n)e^{j2\pi\nu_k t} \, dt \quad (5.31)$$

Then, by using Equation (5.29), $\chi_n^s(\tau,0)$ can be rewritten as follows:

$$\chi_n^{(s)}(\tau,0) = \sigma_k e^{j2\pi\nu_k nT_B} e^{j2\pi\nu_k\tau_k} A_n(\tau-\tau_k,-\nu_k) \qquad (5.32)$$

Finally, by using Equation (5.17), $\chi_b^s(\tau, \nu)$ can be derived as follows:

$$\chi_b^{(s)}(\tau, \nu) = \sigma_k e^{-j2\pi\nu T_B} e^{j2\pi\nu_k \tau_k} \sum_{n=0}^{n_B-1} e^{-j2\pi(\nu-\nu_k)n T_B} A_n(\tau - \tau_k, -\nu_k) \quad (5.33)$$

The noise components, $\chi_b^{(w)}(\tau, \nu)$ and $\chi_{opt}^{(w)}(\tau, \nu)$, are computed by substituting the noise signal in the nth batch, $w(t, n)$, in Equation (5.11) and (5.15), respectively.

Let us consider $w(t, n)$ a stochastic process, then, by applying Equation (5.11), $\chi_{opt}^{(w)}(\tau, \nu)$ can be computed as follows:

$$\chi_{opt}^{(w)}(\tau, \nu) = \sum_{n=0}^{n_B-1} e^{-j2\pi\nu n T_B} \chi_n^{(w)}(\tau, \nu) \quad (5.34)$$

The mean power of $\chi_{opt}^{(w)}(\tau, \nu)$ can be computed as in Equation (5.35)

$$E\left\{|\chi_{opt}^{(w)}(\tau, \nu)|^2\right\} = \sum_n \sum_l e^{-j2\pi(n-l)\nu T_B} E\left\{\chi_n^{(w)}(\tau, \nu) \cdot \left(\chi_l^{(w)}(\tau, \nu)\right)^*\right\} \quad (5.35)$$

where $(\cdot)^*$ denotes the complex conjugate operator. By assuming that the noise in different batches is uncorrelated, that is $E\left\{\chi_n^{(w)}(\tau, \nu) \cdot \left(\chi_l^{(w)}(\tau, \nu)\right)^*\right\} = 0$ when $n \neq l$, Equation (5.35) can be rewritten as follows:

$$E\left\{|\chi_{opt}^{(w)}(\tau, \nu)|^2\right\} = \sum_n E\left\{|\chi_n^{(w)}(\tau, \nu)|^2\right\} \quad (5.36)$$

By looking at Equation (5.11), the integral can be interpreted as the output of a filter whose impulse response is

$$h_{FA_{opt}}(t, n) = s_{refb}(t, n) e^{-j2\pi\nu t} \quad (5.37)$$

Under the assumption of wide-sense stationary random process, the ESD of the noise signal at the output of the filter is as follows:

$$\mathcal{P}_{n_{opt}}(f, n) = \mathcal{P}_w(f, n) \cdot |H_{FA_{opt}}(f, n)|^2 \quad (5.38)$$

where $\mathcal{P}_w(f, n)$ is the ESD of $w(t, n)$, $H_{FA_{opt}}(f, n)$ is the Fourier transform with respect to the variable t of $h_{FA_{opt}}(t, n)$ and is defined as follows:

$$H_{FA_{opt}}(f, n) = S_{refb}(f - \nu, n) \quad (5.39)$$

Then, the mean power of $\chi_n^{(w)}(\tau, \nu)$ can be obtained as in Equation (5.40).

$$E\left\{|\chi_n^{(w)}(\tau, \nu)|^2\right\} = \int \mathcal{P}_w(f, n) \cdot |S_{refb}(f - \nu, n)|^2 \, df \quad (5.40)$$

If we consider Equation (5.15) and we look at the definition of $\chi_{cc}(\tau, n)$, similar considerations can be drawn, but in this case the filter can be defined as follows:

$$h_{FA_b}(t, n) = s_{refb}(t, n) \tag{5.41}$$

and

$$\mathcal{P}_{n_b}(f, n) = \mathcal{P}_w(f, n) \cdot |S_{refb}(f, n)|^2 \tag{5.42}$$

Then, the mean power of $\chi_b^{(w)}(\tau, \nu)$ can be derived as follows:

$$E\left\{|\chi_b^{(w)}(\tau, \nu)|^2\right\} = \sum_n E\left\{|\chi_{cc}^{(w)}(\tau, n)|^2\right\} \tag{5.43}$$

where

$$E\left\{|\chi_{cc}^{(w)}(\tau, n)|^2\right\} = \int \mathcal{P}_w(f, n) \cdot |S_{refb}(f, n)|^2 \; df \tag{5.44}$$

Under the assumption of Gaussian and white noise, the noise at the output of both the batch algorithm and the optimum algorithm have the same mean power level,

$$E\left\{|\chi_n^{(w)}(\tau, \nu)|^2\right\} = E\left\{|\chi_{cc}^{(w)}(\tau, n)|^2\right\} \tag{5.45}$$

then,

$$E\left\{|\chi_{opt}^{(w)}(\tau, \nu)|^2\right\} = E\left\{|\chi_b^{(w)}(\tau, \nu)|^2\right\} \tag{5.46}$$

Even in the hypothesis of colored noise, but still uncorrelated noise realizations among the batches, Equation (5.46) holds true when the maximum Doppler frequency of the target ν_{max} is much smaller than the bandwidth of the reference signal, which is usually the case.

By using Equation (5.33) and Equation (5.30) under the hypothesis of a single slowly fluctuating point target, the $IF(\tau, \nu)$ can be approximated as follows:

$$IF(\tau, \nu) = 20 log_{10}\left(\frac{|\sum_{n=0}^{n_B-1} e^{-j2\pi(\nu-\nu_k)nT_B} A_n(\tau - \tau_k, -\nu_k)|}{|\sum_{n=0}^{n_B-1} e^{-j2\pi(\nu-\nu_k)nT_B} A_n(\tau - \tau_k, \nu - \nu_k)|}\right) \tag{5.47}$$

where τ_k and ν_k are the delay-time and Doppler shift of a single slowly fluctuating point target. Equation (5.47) shows how the losses only depend on the target Doppler frequency and on the shape of the ambiguity function $A_n(0, \nu)$, evaluated within each batch, along the Doppler coordinate. Specifically, if the small Doppler approximation can be accepted, the losses can be considered negligible. Moreover, the small Doppler approximation is acceptable if the product $\nu_{\max}T_B$ is small and for this reason, the batch length can be shortened as required in order to obtain reduced losses. It should be mentioned that the performance will depend on the type of waveform, as some waveforms are intrinsically less influenced by the Doppler effect than others. The computational load of the batch algorithm increases when the number of batches, n_B, and the number of range bins in the CAF increase. In other words, if the batch length T_B decreases, the number of batches n_B increases and consequently the computational effort results are more expensive due to the higher number of cross-correlation functions to be computed. In conclusion, the batch length can be set in order to obtain a good compromise between the SNR losses and the computational cost.

5.3   PASSIVE ISAR THEORY

Before introducing the passive ISAR concept, we should discuss its feasibility, that is, whether a PR system meets the requirements needed to obtain effective ISAR images. These requirements are

- Phase coherence that should be kept high during the coherent processing interval (CPI).
- High enough pulse repetition frequency (PRF) to avoid image aliasing.
- Fine range and cross-range resolutions.

Phase coherence is ensured by the cross-correlation between the direct and the surveillance signals.

Since IOs transmit continuously, the required PRF to avoid image aliasing can be set according to the scenario characteristic. In a passive radar, in fact, the PRF coincides with the batch length, namely T_B. Then, it is possible to set it to avoid image aliasing and at the same time to minimize SNR losses, as already said. Differently from an active radar system which transmits the same waveform every PRI, in a passive radar system, the transmitted signal may change among different batches.

Finally, concerning the spatial resolutions, it must be pointed out that the signal waveform, and specifically the central frequency and the transmitted instantaneous bandwidth, as well as the bistatic geometry strongly affect the range and cross-range resolutions. Although these factors are not under the control of the radar system designer, the available waveform parameters can be properly managed to obtain fine enough spatial resolutions, as it will be shown in the following sections.

5.3.1   PASSIVE ISAR SIGNAL PROCESSING CHAIN

The passive ISAR algorithm takes as input the RD map which may contain more than one target. However, the ISAR image processing can process a single target at time, therefore, the signal scattered by the target of interest

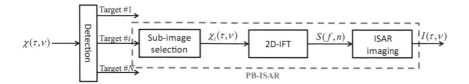

Figure 5.4   Passive bistatic ISAR block scheme

must be extracted before. This operation cannot be performed in the data domain as the received signal is the sum of the echoes from all the targets in the scene. It is carried out in the range-Doppler domain (τ, ν), where the moving targets, even if unfocused, appear as separated objects.

Once target detection has been performed, a sub-image of the target of interest is extracted by selecting a rectangular window in the RD map, $\chi_i(\tau, \nu)$, which contains only the target echo plus a small amount of clutter and noise. The target sub-image selection acts as a two-dimensional filter and is essential because of the following reasons:

- More than one target may be present in the RD map, but only a target at a time can be processed by the ISAR processing.
- Noise and clutter are strongly attenuated as only the resolution cells occupied by the target are retained whereas the rest is filtered out.

The sub-image of the target of interest is then passed to the ISAR imaging algorithm. ISAR processors are typically designed to work in the frequency/slow-time domain, as shown in Figure 5.4, where the variable f represents the frequency while the discrete variable n represents the slow-time. Therefore, the target sub-image should be suitably transformed into the above-mentioned domain. This operation is carried out by means of a two-dimensional Fourier transform. Once the signal scattered by the target of interest in the data domain has been obtained it can be processed by the ISAR algorithm which consists of both autofocus and image formation.

The mathematical details of both the target extraction steps and the ISAR image formation steps will be given in the next subsections.

5.3.2   P-ISAR SIGNAL MODELING

Let the geometry be represented by Figure 5.5, where $T_x(x_1, x_2, x_3)$ is a Cartesian reference system embedded on the target so that x_2 is aligned with the bistatic LoS, $i_{LoS,B}$ at the central time of the observation time interval, x_3 is aligned with the bistatic target effective rotation vector, $\mathbf{\Omega}_B$ (the projection of the target bistatic total rotation vector $\mathbf{\Omega}_{BT}$ onto the plane perpendicular to $i_{LoS,B}$) and x_1 is chosen so as to be perpendicular to the plane

(x_2, x_3). $T_\xi (\xi_1, \xi_2, \xi_3)$ is a Cartesian reference system centered on the transmitter and $R_{T_x T_g}$, $R_{T_x R_x}$ and $R_{R_x T_g}$ are the transmitter-target, transmitter-receiver and the receiver-target distances. The receiver is composed of two antennas, namely, the reference antenna, A_{ref}, which points towards the transmitter to acquire the reference signal, $s_{ref}(t)$, and the surveillance antenna, A_R, which points towards the target to acquire the surveillance signal, $s_R(t)$.

As already said, a passive radar is intrinsically bistatic, since the transmitter and the receiver are typically not co-located. As shown in [13] a bistatic configuration can be approximated with an equivalent monostatic configuration with a virtual sensor located along the bisector of the bistatic angle β and at a distance $R = (R_{T_x T_g} + R_{R_x T_g})/2$, which is the semi-sum of the distances $R_{R_x T_g}$ and $R_{T_x T_g}$. Therefore, a bistatic equivalent monostatic (BEM) radar can be defined as shown in Figure 5.5, and the bistatic ISAR theory can be applied in this framework.

The target angular motion with respect to the BEM radar can be described by means of the total target angular rotation vector $\mathbf{\Omega}_{BT}$. The effective rotation vector $\mathbf{\Omega}_B$ is the rotation vector component that contributes to the synthetic aperture formation. This latter can be obtained from the total target rotation vector as follows:

$$\mathbf{\Omega}_B = \boldsymbol{i}_{LoS,B} \times [\mathbf{\Omega}_{BT} \times \boldsymbol{i}_{LoS,B}] \tag{5.48}$$

where $\boldsymbol{i}_{LoS,B}$ is the unit vector which identifies the BEM radar line of sight and $\times$ represents the cross-product operator.

As shown in the previous section, the cross-ambiguity function between the surveillance signal, $s_R(t)$, and the reference signal, $s_{ref}(t)$, can be seen as a weighted sum of the ambiguity functions calculated within each batch. Specifically, Equation (5.16) highlights that the cross-ambiguity function, $\chi_b(\tau, \nu)$, is the Fourier transform of the sequence $\chi_{cc}(\tau, \nu)$. Moreover, as previously stated, since the ISAR processing can process a single target at a time, a sub-image of the ith target, $\chi_i(\tau, \nu)$, is cropped out from the RD map prior to the application of the ISAR algorithm [12].

Therefore, by inverse Fourier transforming $\chi_i(\tau, \nu)$ with respect to the variable ν, the range profile history, $\chi_{cc,i}(\tau, n)$ of the ith target is derived:

$$\chi_{cc,i}(\tau, n) = FT^{-1}[\chi_i(\tau, \nu)] \tag{5.49}$$

Let us now see how $\chi_{cc,i}(\tau, n)$ is related with the signal scattered by the ith target. For the sake of simplicity, in the following equations the index i has been eliminated. The signal scattered by the target can be written as follows:

$$s_R(t) = \int_V f'(\mathbf{x}) s_{ref}(t - \tau(t; \mathbf{x})) \, d\mathbf{x} \tag{5.50}$$

where, $\tau(t; \mathbf{x}) = \frac{R(\mathbf{x}, t)}{c}$, $R(\mathbf{x}, t) = \frac{R_{T_x T_g}(\mathbf{x}, t) + R_{R_x T_g}(\mathbf{x}, t)}{2}$ is the bistatic distance between an arbitrary point on the target, the receiver and the transmitter, $\mathbf{x}$

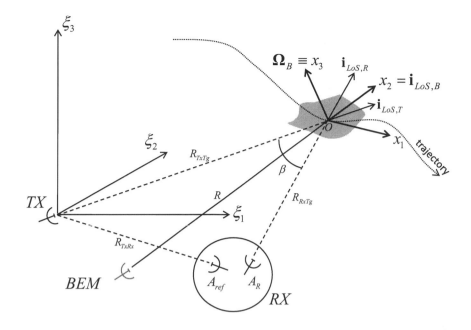

Figure 5.5 Geometry

is the vector locating the point of the target in the coordinates system T_x, c is the light speed in a vacuum and V is the spatial domain where the target reflectivity function, $f'(\mathbf{x})$, is defined.

By assuming the target stationary during the batch interval, T_B, the delay-time can be approximated as $\tau(\mathbf{x}, t) \simeq \tau(\mathbf{x}, n)$. Therefore, by using Equation (5.2), Equation (5.50) can be reformulated as follows:

$$s_R(t) = \sum_{n=0}^{n_B-1} \int_V f'(\mathbf{x}) s_{refb}\left(t - \tau(\mathbf{x}, n) - nT_B, n\right) \ d\mathbf{x} \qquad (5.51)$$

As mentioned above, in order to have the same integration gain at each batch, the surveillance signal is divided into segments of length $T_B + \tau_{max}$. According to that, the received signal, $s_R(t)$ in the nth segment can be written as in the following formula:

$$s_{Rb}(t, n) = \int_V f'(\mathbf{x}) \left[s_{refb}\left(t - \tau(\mathbf{x}, n) - nT_B, n\right) + s_A(t, n; \mathbf{x}) + s_B(t, n; \mathbf{x})\right]$$

$$(5.52)$$

where $s_{refb}(t - \tau(\mathbf{x}, n) - nT_b, n)$ represents the desired received signal, whereas $s_A(t, n; \mathbf{x})$ and $s_B(t, n; \mathbf{x})$ are portions of the reference signal in the signal segments adjacent to the nth one, as graphically shown in Figure 5.3, namely:

$$s_A(t, n; \mathbf{x}) =$$

$$s_{ref}(t - \tau(\mathbf{x}, n) - (n-1)T_B, n-1) \cdot rect\left(\frac{t - \tau(\mathbf{x}, n)/2}{\tau(\mathbf{x}, n)}\right) \qquad (5.53)$$

$$s_B(t, n; \mathbf{x}) =$$

$$s_{ref}(t - \tau(\mathbf{x}, n) - (n+1)T_B, n+1) \cdot rect\left(\frac{t - \left(T_b + \frac{\tau_{max} + \tau(\mathbf{x}, n)}{2}\right)}{\tau_{max} - \tau(\mathbf{x}, n)}\right)$$

$$(5.54)$$

By using Equation (5.52), the cross-ambiguity function in a segment is:

$$\chi_{cc}(\tau, n) = \chi_{cc}^{(s)}(\tau, n) + \chi_A(\tau, n) + \chi_B(\tau, n) \qquad (5.55)$$

where

$$\chi_A(\tau, n) = \int_V f'(\mathbf{x}) \int_{t=-\infty}^{\infty} s_A(t, n; \mathbf{x}) s_{refb}^*(t - \tau, n) \, dt \, d\mathbf{x} \qquad (5.56)$$

$$\chi_B(\tau, n) = \int_V f'(\mathbf{x}) \int_{t=-\infty}^{\infty} s_B(t, n; \mathbf{x}) s_{refb}^*(t - \tau, n) \, dt \, d\mathbf{x} \qquad (5.57)$$

$$\chi_{cc}^s(\tau, n) = \int_V f'(\mathbf{x}) \int_{t=-\infty}^{\infty} s_{refb}(t - \tau(\mathbf{x}, n), n) s_{refb}^*(t - \tau, n) \, dt \, d\mathbf{x} \quad (5.58)$$

Equation (5.55) can be approximated as follows:

$$\chi_{cc}(\tau, n) \simeq \chi_{cc}^{(s)}(\tau, n) \qquad (5.59)$$

In fact, the term $\chi_{cc}^{(s)}(\tau, n)$ is larger than $\chi_A(\tau, n)$ and $\chi_B(\tau, n)$, since the signals in different batches are typically uncorrelated. In any case, even when a small correlation exists, both $\chi_A(\tau, n)$ and $\chi_B(\tau, n)$ can be neglected as their averages are statistically much smaller than the average of $\chi_{cc}^{(s)}(\tau, n)$.

Equation (5.58) through Equation (5.59) express the relationship between the target received signal and the related cross-ambiguity function. Let us now project $\chi_{cc}(\tau, \nu)$ into the frequency/slow-time domain.

By Fourier transforming $\chi_{cc}(\tau, n)$ with respect to the variable τ, the signal $S(f, n)$ in the frequency/slow-time domain is obtained as follows:

$$
\begin{aligned}
S(f,n) &= FT\left[\chi_{cc}(\tau,n)\right] \\
&= \int_V f'(\mathbf{x}) \int_t s_{refb}(t-\tau(\mathbf{x},n),n) \int_\tau s^*_{refb}(t-\tau,n)e^{-j2\pi f\tau} \ d\tau \, dt \, d\mathbf{x} \\
&= \int_V f'(\mathbf{x}) \int_t s_{refb}(t-\tau(\mathbf{x},n),n)S^*_{refb}(f,n)e^{-j2\pi ft} \ dt \, d\mathbf{x} \\
&= \int_V f'(\mathbf{x})S_{refb}(f,n)S^*_{refb}(f,n)e^{-j2\pi f\tau(\mathbf{x},n)} \ d\mathbf{x} \\
&= |S_{refb}(f,n)|^2 \int_V f'(\mathbf{x})e^{-j2\pi f\tau(\mathbf{x},n)} \ d\mathbf{x}
\end{aligned}
$$

$$(5.60)$$

where $S_{refb}(f,n)$ is the Fourier transform of $s_{refb}(\tau,n)$.

The delay-time relative to an arbitrary scatterer can be written as follows:

$$
\tau(\mathbf{x},n) = \frac{R_{TxTg}(\mathbf{x},n) + R_{RxTg}(\mathbf{x},n)}{c} \tag{5.61}
$$

where $R_{TxTg}(\mathbf{x},n)$ and $R_{RxTg}(\mathbf{x},n)$ are the distances between an arbitrary scatterer and the transmitter and the receiver, respectively. When the target size is much smaller than both the target-transmitter and the target-receiver distances, the straight-iso-range approximation can be applied, therefore both $R_{TxTg}(\mathbf{x},n)$ and $R_{RxTg}(\mathbf{x},n)$ can be approximated as:

$$
\begin{aligned}
R_{TxTg}(\mathbf{x},n) &= R_{TxTg}(n) + \mathbf{x} \cdot \mathbf{i}_{LoS,T}(n) \\
R_{RxTg}(\mathbf{x},n) &= R_{RxTg}(n) + \mathbf{x} \cdot \mathbf{i}_{LoS,R}(n)
\end{aligned} \tag{5.62}
$$

where $R_{TxTg}(n)$ and $R_{RxTg}(n)$ are the distances between the focusing point O and the transmitter and receiver, respectively, and $\mathbf{i}_{LoS,T}$ and $\mathbf{i}_{LoS,R}$ are the unit vectors which identify the LoS of the transmitter and the receiver, respectively (see Figure 5.5).

Therefore, by using Equation (5.62), Equation (5.61) can be reformulated as follows:

$$
\tau(\mathbf{x},n) = \frac{2}{c}\left[R(n) + K(n)\mathbf{x} \cdot \mathbf{i}_{LoS,B}(n)\right] \tag{5.63}
$$

where:

$$
R(n) = \frac{R_{TxTg}(n) + R_{RxTg}(n)}{2}
$$

$$
K(n) = \left|\frac{\mathbf{i}_{LoS,T}(n) + \mathbf{i}_{LoS,R}(n)}{2}\right| = cos\left(\frac{\beta(n)}{2}\right) \tag{5.64}
$$

$$
\mathbf{i}_{LoS,B}(n) = \frac{\mathbf{i}_{LoS,T}(n) + \mathbf{i}_{LoS,R}(n)}{|\mathbf{i}_{LoS,T}(n) + \mathbf{i}_{LoS,R}(n)|}
$$

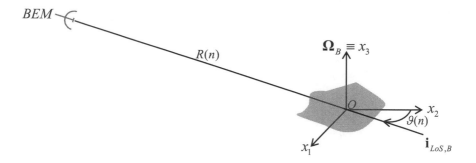

Figure 5.6 Pictorial representation of the aspect angle $\vartheta(n)$. The aspect angle is defined as the angle between the radar LoS, in this case $\mathbf{i}_{LoS,B}$ and x_2

where $\beta(n)$ represents the value of the bistatic angle at time n (see Figure 5.5 for the geometrical interpretation of β).

Then, Equation (5.60) can be rewritten as follows:

$$S(f,n) = |S_{refn}(f,n)|^2 e^{-j4\pi\frac{f}{c}R(n)} \int_V f'(\mathbf{x}) e^{-j4\pi\frac{f}{c}K(n)\mathbf{x}\cdot\mathbf{i}_{LoS,B}(n)} \ d\mathbf{x} \quad (5.65)$$

$S(f,n)$ in Equation (5.65) represents the signal model in the frequency/slow-time domain to which the imaging algorithm is applied.

5.3.3 P-ISAR IMAGING

The P-ISAR imaging algorithm proposed in this book is based on the image contrast based autofocus (ICBA) algorithm and the range-Doppler technique. The ICBA is a parametric technique and is based on the image contrast (IC) maximization. Details can be found in Chapter 3.

The range-Doppler technique is based on the assumption that the Doppler frequency of each scatterer, relative to a reference point taken on the target, is constant within the observation time. This assumption is usually verified when the *effective rotation vector* $\mathbf{\Omega}_B(t)$ can be assumed constant within the observation time. Let $\vartheta(n)$ be the target aspect angle at the slow-time instant n with respect to the LoS of the *BEM* radar, as shown in Figure 5.6. The scalar product $\mathbf{x} \cdot \mathbf{i}_{LoS,B}(n)$ can be reformulated as in Equation (5.66):

$$\mathbf{x} \cdot \mathbf{i}_{LoS,B}(n) = x_1 sin(\vartheta(n)) + x_2 cos(\vartheta(n)) \quad (5.66)$$

where x_1 and x_2 are the scatterer coordinates in the image projection plane (IPP), which is the plane orthogonal to $\mathbf{\Omega}_B$. For notation simplicity, and as it can be noted in Figure 5.5, the reference system embedded on the target,

T_x, has been chosen so that the two-dimensional plane (x_1, x_2) coincides with the ISAR image projection plane. Then, $S(f, n)$ can be rewritten as:

$$S(f, n) = |S_{refb}(f, n)|^2 e^{-j4\pi \frac{f}{c} R(n)} \int_V f'(\mathbf{x}) e^{-j2\pi [x_1 X_1(f,n) + x_2 X_2(f,n)]} \, d\mathbf{x}$$

(5.67)

where

$$\begin{aligned} X_1(f, n) &= \frac{2f K(n) \sin(\vartheta(n))}{c} \\ X_2(f, n) &= \frac{2f K(n) \cos(\vartheta(n))}{c} \end{aligned}$$

(5.68)

are the spatial frequencies.

The variation of the distance between the target and the BEM, namely $R(n)$, over the slow-time domain, causes a range migration that must be accounted for. In order to compensate the target radial motion, the phase term $e^{-j4\pi \frac{f}{c} R(n)}$ in Equation (5.68) must be estimated and removed. In ISAR scenarios, where the target is usually non-cooperative, this operation is performed by means of autofocusing algorithms. The estimation of $e^{-j4\pi \frac{f}{c} R(n)}$ results in the estimation of $R(n)$ and therefore of the target motion parameters.

After target motion compensation and some mathematical manipulations, the signal in Equation (5.67) can be expressed as in Equation (5.69)

$$\begin{aligned} S_C(f, n) &= W(f, n) \int_{x_1} \int_{x_2} f(x_1, x_2) e^{-j2\pi [x_1 X_1(f,n) + x_2 X_2(f,n)]} \, dx_1 \, dx_2 \\ &= W(f, n) \cdot G(X_1, X_2) \end{aligned}$$

(5.69)

where the dependency from (f, t) has been omitted in X_1 and X_2 and $W(f, n)$ represents the signal support of $S(f, n)$, which is the region in the frequency/slow-time domain where the signal $S(f, n)$ exists, and is expressed as follows:

$$W(f, n) = |S_{refb}(f, n)|^2 W_T(n)$$

(5.70)

where $W_T(n) = rect\left(\frac{n}{N}\right)$, $N = \lceil \frac{T_{int}}{T_B} \rceil$, $f(x_1, x_2) = \int_{x_3} f'(\mathbf{x}) \, dx_3$, is the projection of the target reflectivity function onto the IPP, and $F(X_1, X_2) = 2D\text{-}FT[f(x_1, x_2)]$, is the two-dimensional Fourier transform of $f(x_1, x_2)$. As can be noted by observing Equation (5.69), an estimate of $f'(x_1, x_2)$ can be obtained by inverse Fourier transforming $S_C(f, n)$.

However, to effectively apply the range-Doppler algorithm, the signal support band, $W(f, n)$ shall be rectangular, otherwise an interpolation algorithm should be used before the ISAR image reconstruction.

Then, under the hypothesis that the aspect angle variation within the integration time is small enough $|\Delta\vartheta| \ll 1$, and that the bistatic angle changes are relatively small, $\beta(n) \simeq \beta(0) = \beta_0$, both the spatial frequencies can be approximated as follows:

$$X_2(f, n) \simeq X_2(f) = \frac{2fK_0}{c}$$

$$X_1(f, n) \simeq X_1(n) = \frac{2f_0K_0\vartheta(n)}{c} = \frac{2f_0K_0\Omega_B nT_B}{c} \tag{5.71}$$

where $K_0 = \frac{\cos(\beta(0))}{2}$.

By substituting both $X_1(f, n)$ and $X_2(f, n)$ in Equation (5.69), the motion compensated signal becomes:

$$S_C(f, n) = W(f, n) \int_{x_1} \int_{x_2} f(x_1, x_2) e^{-j2\pi \left[x_1 \frac{2fK_0}{c} + x_2 \frac{2f_0K_0\Omega_B nT_B}{c} \right]} dx_1 \, dx_2 \tag{5.72}$$

By substituting the variables (x_1, x_2) with the variables (τ, ν) as defined in Equation (5.73)

$$\nu = \frac{2f_0K_0\Omega_B x_1}{c}$$

$$\tau = \frac{2x_2K_0}{c} \tag{5.73}$$

Equation (5.72) can be rewritten as follows:

$$S_C(f, n) = C \cdot W(f, n) \int_\tau \int_\nu f(\tau, \nu) e^{-j2\pi[\tau f + \nu nT_B]} \, d\tau \, d\nu \tag{5.74}$$

where $C = \frac{c^2}{f_0 K_0^2 \Omega_B}$.

By observing Equation (5.74), it can be noted that both range and Doppler resolutions are degraded by a factor K_0 with respect to those achievable with a monostatic radar.

By applying the RD technique to the signal in Equation (5.74), the ISAR image of the target is obtained:

$$I(\tau, \nu) = Cw(\tau, \nu) \otimes \otimes f(\tau, \nu) \tag{5.75}$$

where $w(\tau, \nu)$ represents the point spread function (PSF) of the ISAR system in the delay-time/Doppler domain, which depends on both the spectral content of the reference signal and on the integration time. The symbol $\otimes\otimes$ indicates double convolution.

It is important to remark that the results obtained in this section are independent of the illuminator of opportunity and therefore independent of the reference signal waveform.

In the following section, the ISAR system performance in terms of spatial resolutions will be analyzed.

5.3.4 PERFORMANCES ANALYSIS

ISAR system performances are usually evaluated in terms of spatial resolutions, which can be calculated through the analysis of the system PSF. The purpose of this section is to find a closed form solution of the system PSF.

In passive radars, the reference signal in a batch, $s_{refb}(t, n)$, can be modeled as a random signal, where the index n indicates a realization of the stochastic process. Differently from active radars, in passive radars, the signal content usually changes over time. Then, differently from active radars in which the transmitted waveform is unchanged over the slow-time, in the passive radar case, the reference signal changes over the slow-time variable. Because of that, only the statistical average of the spatial resolutions can be computed.

As shown in Equation (5.74), the signal support $W(f, n)$ is proportional to $|S_{refb}(f, n)|^2$, which is the ESD of $s_{refb}(t, n)$ which is a realization of a random signal. The ESD can be computed then as the statistical average of $|S_{refb}(f, n)|^2$:

$$\hat{\mathcal{P}}_{S_{ref}}(f) = \frac{E\left[|S_{refb}(f, n)|^2\right]}{T_B} \tag{5.76}$$

Equation (5.76) is however an approximation of $\mathcal{P}_{S_{ref}}(f)$, being $\mathcal{P}_{S_{ref}} = \lim_{T_B \to \infty} \frac{E\left[|S_{refb}(f,n)|^2\right]}{T_B}$.

However, $\hat{\mathcal{P}}_{S_{ref}}(f)$ is typically a consistent estimator of $\mathcal{P}_{S_{ref}}(f)$.

Then, the signal support band can be written as a function of $\hat{\mathcal{P}}_{S_{ref}}(f)$:

$$\overline{W}(f, n) = \hat{\mathcal{P}}_{S_{ref}}(f) \cdot W_T(n) \tag{5.77}$$

By Fourier transforming Equation (5.77), the statistical average of the PSF can be derived:

$$\overline{w}(\tau, \nu) = \hat{p}_{ref}(\tau) \cdot w_T(\nu) \tag{5.78}$$

where $w_T(\nu)$ is the Fourier series of $W_T(n)$:

$$w_T(\nu) = e^{j\pi(N-2)\nu T_B} \frac{sin(n_B T_B \pi \nu)}{sin(\pi \nu T_B)} \tag{5.79}$$

It is worth pointing out that the result in Equation (5.78) is a direct consequence of the application of a two-dimensional Fourier transform to a function which is the product of separate functions of different variables.

Moreover, it should be pointed out that the amplitude of $w_T(\nu)$ approaches that of the function $N \cdot sinc(\nu n_B T_B)$. By observing Equation (5.79) the shape of the PSF along the Doppler coordinate is independent of the reference signal content, and, therefore, the Doppler resolution is determined by the integration time $T_{obs} = n_B T_B$ only. Instead, the characteristics of $\hat{p}_{ref}(\tau)$ affect the PSF along the time-delay coordinate.

5.4 ILLUMINATORS OF OPPORTUNITY ANALYSIS

The illuminators of opportunity can be divided into two main classes: analogue emitters like frequency modulation (FM) radio [9], analogue television, or digital transmitter as digital audio broadcasting (DAB) [8] and digital video broadcasting-terrestrial (DVB-T) [5, 4], global system for mobile communications (GSM), universal mobile telecommunications systems (UTMS) [21], WiFi and worldwide interoperability for microwave access (WiMAX). The parameters that need to be taken into account in assessing their usefulness are: their power density at the target, their coverage (both spatial and temporal) and the signal characteristics. In general, digital waveforms have to be preferred to analog ones, since the ambiguity function is independent on the signal content. Passive radars based on analogue signals show detection performance strongly dependent on the signal content. In contrast, digital waveforms, thanks to specific signal coding, have spectral properties which are nearly independent of the signal content. Such waveforms exhibit an ambiguity function with a thumb-tack shape and a bandwidth that is constant in time. In Table 5.1, the main parameters of different types of waveforms are reported.

Table 5.1

Summary of typical parameters of PBR IOs

IOs	Frequency [MHz]	EIRP [KW]	Istantaneous Bandwidth [MHz]	Monostatic Range Resolution [m]
FM	87.5-108	0.01-0.1	0.16	937
UMTS	2210-2170	0.001-0.1	3.84	39
DAB	174-240	0.8-1.6	1.536	100
DVB-T	164-800	0.1-10	7.61	19.7
WiFi 802.11	2400	0.0001	5	30
Wi-MAX	2400	0.02	20	15

Among digital waveforms, DVB-T, DAB, GSM and recently UMTS are the most exploited as illuminator of opportunity. Specifically, both UMTS and DVB-T signals have a wide bandwidth that allows achieving good spatial resolution. Among digital waveform the best compromise between range resolution and EIRP value can be achieved by using DVB-T signals.

5.4.1 DVB-T CASE STUDY

In this section we derive the statistical average of the PSF when a DVB-T transmitter is used as an illuminator of opportunity. Among all digital IOs, a DVB-T signal exhibits a content-independent spectrum and a relatively wide bandwidth. The DVB-T signal is organized in COFDM frames. Each frame consists of 68 OFDM symbols. Each symbol is formed by a set of data sub-carriers, pilot sub-carriers and transport parameter signaling (TPS). The pilot and TPS sub-carriers are used for receiver synchronization and transmission parameter estimation, respectively. In the Italian DVB-T standard, 6817 sub-carriers, among the available 7168, are used to transmit. This means that the signal bandwidth is $B_S = 7.61\,MHz$ while the canalization is equal to $B_C = 8\,MHz$. Because of that, adjacent DVB-T channels are separated by gaps where no signal is transmitted. Such gaps are useful in order to reduce the cross-interference between adjacent channels. Each sub-carrier is 64 QAM modulated with baseband data.

As already said, the reference signal affects only the PSF along the delay-time domain, whereas the PSF along the Doppler domain is determined only by the integration time and is given in Equation (5.79).

The delay-time component of the PSF evaluated at zero-Doppler ($\nu = 0$) is obtained here by using real data. The $\hat{\mathcal{P}}_{S_{ref}}(f)$ is first obtained by applying Equation (5.76) to the acquired data, then the delay-time component of the PSF at zero-Doppler is calculated by applying a Fourier transform to $\hat{\mathcal{P}}_{S_{ref}}(f)$.

It should be noted that for the case at hand, the delay-time component of the PSF remains the same for all Doppler frequency except for an amplitude value ($w_T(\nu)$). Specifically, the case of both single DVB-T channel and multiple adjacent DVB-T channels will be considered hereinafter with the aim of analyzing the system PSF. Experimental data was acquired by using a software defined radio acquisition board. The equipment used to acquire the DVB-T signal is fully described in [21, 1].

A single channel DVB-T signal was acquired for an observation time equal to $0.15\,s$ by pointing the antenna towards the illuminator of opportunity. In order to compute $\hat{\mathcal{P}}_{S_{ref}}(f)$ the signal has been divided into n_B batches of shorter length. By considering the n_B signals as random realizations of a stochastic process, Equation (5.76) can be applied. In Figure 5.7 the PSD of a single batch signal and the average PSD, namely $\hat{\mathcal{P}}_{S_{ref}}(f)$, are represented. By looking at Figure 5.7 a suitable model that approximates $\hat{\mathcal{P}}_{S_{ref}}(f)$ is as follows:

$$\hat{\mathcal{P}}_{S_{ref}}(f) \simeq \mathcal{P}_{S_{ref}}(f) = rect\left(\frac{f}{B_S}\right) \tag{5.80}$$

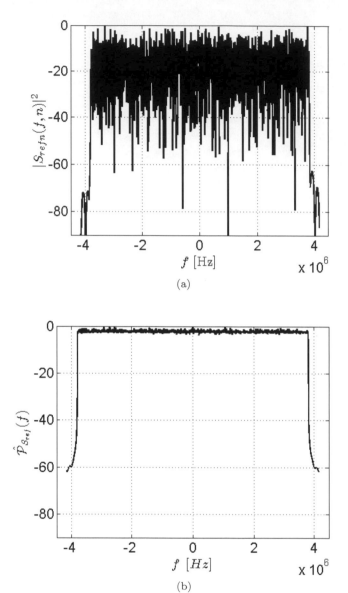

Figure 5.7 (a) PSD of a single batch signal and (b) average PSD, $\hat{\mathcal{P}}_{S_{ref}}(f)$ obtained with $n_B = 700$

Therefore, a model for the delay time component of the PSF can be obtained by Fourier transforming Equation (5.80) as follows:

$$p_{ref}(\tau) = B_S \cdot sinc(\tau B_S) \tag{5.81}$$

$p_{ref}(\tau)$, $\hat{p}_{ref}(\tau)$ and $p_{ref}(\tau, n)$ are represented in Figure 5.8, where $p_{ref}(\tau, n)$ and $\hat{p}_{ref}(\tau)$ have been obtained by applying the Fourier transform to the spectrum in Figure 5.7 (a) and (b), respectively. As it can be observed, $\hat{p}_{ref}(\tau) \simeq p_{ref}(\tau)$.

As recently demonstrated in [4, 5], a wideband signal of opportunity can be obtained by coherently adjoining N_C adjacent DVB-T channels, which produces a range resolution improvement by a factor of N_C.

The use of multiple adjacent DVB-T channels leads to a fine range resolution without drastically affecting the image quality. Some artifacts, however, occur as the gaps between adjacent channels generate grating lobes in the image domain. Some algorithms can be found in literature to solve this problem [17, 25].

A DVB-T signal composed of three adjacent channels was acquired for an observation time equal to $0.15\,s$, by pointing the antenna directly towards the illuminator of opportunity. As it has been done previously, in order to compute $\hat{\mathcal{P}}_{S_{ref}}(f)$, the signal is broken down into n_B batches of shorter length and $\hat{\mathcal{P}}_{S_{ref}}(f)$ is obtained by using Equation (5.76).

As can be noted by observing Figure 5.9, $\hat{\mathcal{P}}_{S_{ref}}(f)$ approximates a model $\mathcal{P}_{S_{ref}}(f)$ which is defined as follows:

$$\hat{\mathcal{P}}_{ref}^{N_C}(f) \simeq \mathcal{P}_{ref}(f) = rect\left(\frac{f}{N_C \cdot B_C}\right) \cdot \sum_{k=-\infty}^{\infty} rect\left(\frac{f - kB_C}{B_S}\right) \tag{5.82}$$

where N_C is the number of adjacent DVB-T channels. Therefore, a model for the PSF in the delay-time domain can be derived by Fourier transforming $\mathcal{P}_{ref}^{N_C}(f)$, as follows:

$$p_{ref}^{N_C}(\tau) = N_C \cdot \sum_{p} sinc\left(\frac{p \cdot B_S}{B_C}\right) \cdot sinc\left(N_C B_C \left(\tau - \frac{p}{B_C}\right)\right) \tag{5.83}$$

$p_{ref}(\tau)$, $\hat{p}_{ref}(\tau)$ and $p_{ref}(\tau, n)$ are represented in Figure 5.10, where $\hat{p}_{ref}(\tau)$ and $p_{ref}(\tau, n)$ are obtained by applying the Fourier transform to the spectrum in Figure 5.9 (a) and (b), respectively. As can be noted, $\hat{p}_{ref}(\tau)$ approximates quite well the PSF model.

Finally, for the sake of comparison, $\hat{p}_{ref}(\tau)$ obtained both for the single DVB-T channel and three DVB-T channels signal, namely $\hat{p}_{ref}(\tau)$ and $\hat{p}_{ref}^{N_C}(\tau)$ with $N_C = 3$, are compared in Figure 5.11, where the resolution enhancement is clearly visible.

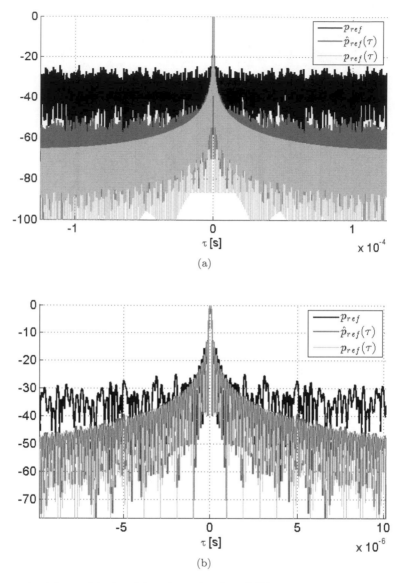

Figure 5.8 (a) $p_{ref}(\tau)$, $\hat{p}_{ref}(\tau)$ and $p_{ref}(\tau, n)$ and (b) zoom

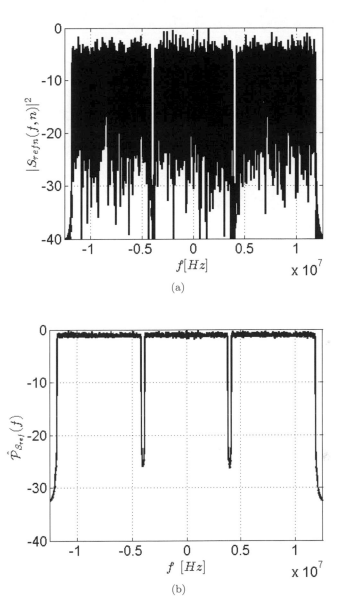

Figure 5.9 (a) PSD of a single batch signal and (b) average PSD, $\hat{\mathcal{P}}_{S_{ref}}(f)$ obtained with $n_B = 700$

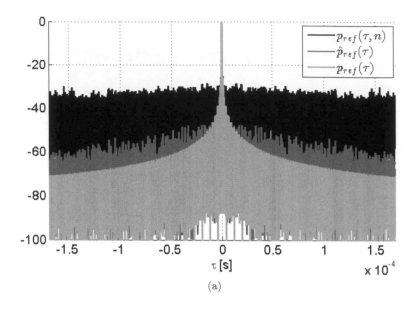

(a)

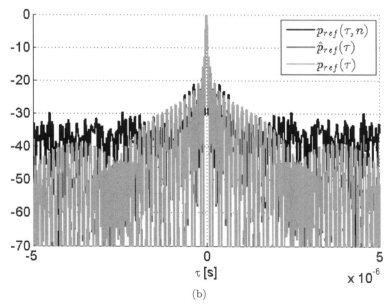

(b)

Figure 5.10 (a) $p_{ref}(\tau)$, $\hat{p}_{ref}(\tau)$ and $p_{ref}(\tau, n)$ and (b) zoom

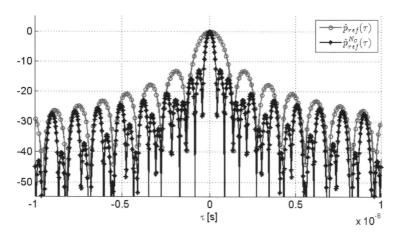

Figure 5.11 Comparison between $\hat{p}_{ref}(\tau)$ and $\hat{p}_{ref}^{N_C}(\tau)$

5.5 CONCLUSIONS

The CAF batch algorithm and the theoretical formulation of P-ISAR imaging, have been provided in this chapter. Moreover, a means to measure the performance of both these algorithms has been provided. Particularly, the performance of the CAF batch algorithm are measured in terms of the improvement factor (IF), that is the ratio of SNR obtained by using the CAF batch algorithm and the SNR obtained by using the conventional CAF method.

The performance of the P-ISAR algorithm are, instead, measured in terms of range and cross-range resolutions. This is because for effective ISAR imaging a mandatory requirement is the achievement of fine range resolution. Differently from active radars, where the transmitted waveforms are properly designed to get the desired spatial resolutions, in a P-ISAR system the spatial resolutions are dependent on the used IO. Specifically, the range resolution is affected by the transmitted instantaneous bandwidth while the cross-range resolution is affected by the operating frequency. The higher the operating frequency, the finer the cross-range resolution.

Even if a particular case study, namely the DVB-T, is handled in this chapter, it should be mentioned that the theoretical results are applicable to all possible IOs, as they are independent of the used waveform.

Finally, it must be pointed out that even if the spatial resolutions in a P-ISAR system are limited by the bandwidth and the operative frequencies, which are typically lower than those of dedicated active radar systems, the joint use of PR systems and P-ISAR processing leads to the formation of bistatic ISAR images of targets at those frequencies where it is usually forbidden to transmit, such as VHF and UHF.

Therefore, P-ISAR may be a useful tool for measuring bistatic RCS of targets of interest, including monostatically stealthy targets.

REFERENCES

1. F. Berizzi, M. Martorella, D. Petri, M. Conti, and A. Capria. USRP technology for multiband passive radar. In *Radar Conference, 2010 IEEE*, pages 225–229, May 2010.
2. M. Cetin and A. D. Lanterman. Region-enhanced passive radar imaging. *Radar, Sonar and Navigation, IEE Proceedings*, 152(3):185–194, 2005.
3. M. Cherniakov. *Bistatic Radars: Emerging Technology*. Wiley, 2008.
4. M. Conti, F. Berizzi, D. Petri, A. Capria, and M. Martorella. High range resolution DVB-T passive radar. In *Proceedings of the European Radar Conference (EURAD)*, 2010.
5. M. Conti, D. Petri, A. Capria, M. Martorella, F. Berizzi, and E. Dalle Mese. Ambiguity function sidelobes mitigation in multichannel DVB-T passive bistatic radar. In *Proceedings of the International Radar Symposium*, 2011.
6. A. Farina and H. Kuschel. *Aerospace and Electronic Systems Magazine, IEEE*, 27(10):5, 2012.
7. J. Garry, C. Baker, G. Smith, and R. Ewing. Doppler imaging for passive bistatic radar. In *IEEE Radar Conference 2013*, 2013.
8. M. Glende, J. Heckenbach, H. Kuschel, S. Muller, J. Schell, and C. Schumacher. Experimental passive radar systems using digital illuminators (DAB/DVB-T). In *Proceedings of the International Radar Symposium*, 2007.
9. P. Howland. Editorial: Passive radar systems. *Radar, Sonar and Navigation, IEE Proceedings*, 152(3):105–106, June 2005.
10. M. Malanowski, G. Mazurek, K. Kulpa, and J. Misiurewicz. FM based PCL radar demonstrator. In *Proceedings of the International Radar Symposium*, 2007.
11. M. Martorella and E. Giusti. Theoretical Foundation of Passive Bistatic ISAR Imaging. *Aerospace and Electronic System, IEEE Transactions on*, 2014.
12. M. Martorella, E. Giusti, F. Berizzi, A. Bacci, and E. Dalle Mese. ISAR based technique for refocussing non-cooperative targets in SAR images. *IET Radar, Sonar, Navigation*, 6(5):1–9, May 2012.
13. M. Martorella, J. Palmer, J. Homer, B. Littleton, and I. D. Longstaff. On bistatic inverse synthetic aperture radar. *IEEE Transactions on Aerospace and Electronic Systems*, 43(3):1125–1134, 2007.
14. W. L. Melvin and J. A. Scheer. *Principles of Modern Radar: Advances Techniques*. Scitech Publishing, 2012.
15. C. Moscardini, D. Petri, M. Conti, M. Martorella, and F. Berizzi. Batches algorithm for passive radar: A theoretical analysis. *Aerospace and Electronic System, IEEE Transactions on*, 2015.
16. D.W. O'Hagan, H. Kuschel, M. Ummenhofer, J. Heckenbach, and J. Schell. A multi-frequency hybrid passive radar concept for medium range air surveillance. *Aerospace and Electronic Systems Magazine, IEEE*, 27(10):6–15, Oct. 2012.

17. D. Olivadese, E. Giusti, D. Petri, M. Martorella, A. Capria, and F. Berizzi. Passive ISAR with DVB-T signals. *Geoscience and Remote Sensing, IEEE Transactions on*, 51(8):4508–4517, Aug. 2013.

18. J. Palmer, S. Palumbo, A. Summers, D. Merrett, S. Searle, and S. Howard. An overview of an illuminator of opportunity passive radar research project and its signal processing research directions. *Digital Signal Processing*, 21(5):593–599, 2011.

19. J.E. Palmer and S.J. Searle. Evaluation of adaptive filter algorithms for clutter cancellation in passive bistatic radar. In *Radar Conference (RADAR), 2012 IEEE*, pages 0493–0498, May 2012.

20. D. Pasculli, A. Baruzzi, C. Moscardini, D. Petri, M. Conti, and M. Martorella. Dvb-t passive radar tracking on real data using extended kalman filter with doa estimation. In *Radar Symposium (IRS), 2013 14th International*, volume 1, pages 184–189, June 2013.

21. D. Petri, F. Berizzi, M. Martorella, E. Dalle Mese, and A. Capria. A software defined UMTS passive radar demonstrator. In *Radar Symposium (IRS), 2010 11th International*, pages 1 –4, June 2010.

22. D. Petri, C. Moscardini, M. Conti, A. Capria, J.E. Palmer, and S.J. Searle. The effects of dvb-t sfn data on passive radar signal processing. In *Radar (Radar), 2013 International Conference on*, pages 280–285, Sept. 2013.

23. D. Petri, C. Moscardini, M. Martorella, M. Conti, A. Capria, and F. Berizzi. Performance analysis of the batches algorithm for range-doppler map formation in passive bistatic radar. In *Radar Systems (Radar 2012), IET International Conference on*, pages 1–4, Oct. 2012.

24. D. Petri, C. Moscardini, M. Martorella, M. Conti, A. Capria, and F. Berizzi. Performance analysis of the batches algorithm for range-doppler map formation in passive bistatic radar. In *Radar Systems (Radar 2012), IET International Conference on*, pages 1–4, Oct. 2012.

25. W. Qiu, E. Giusti, A. Bacci, M. Martorella, F. Berizzi, H.Z. Zhao, and Q. Fu. Compressive sensing for passive isar with dvb-t signal. In *Radar Symposium (IRS), 2013 14th International*, volume 1, pages 113–118, June 2013.

26. S. Stein. Algorithms for ambiguity function processing. *Acoustics, Speech and Signal Processing, IEEE Transactions on*, 29(3):588–599, June 1981.

27. T. Tsao, M. Slamani, P. Varshney, D. Weiner, H. Schwarzlander, and S. Borek. Ambiguity function for a bistatic radar. *Aerospace and Electronic Systems, IEEE Transactions on*, 33(3):1041–1051, July 1997.

6 3D Interferometric ISAR

M. Martorella, D. Staglianò, F. Salvetti, and F. Berizzi

CONTENTS

Inverse synthetic aperture radar (ISAR) has been studied for more than three decades and several demonstrations have proven its effectiveness in forming electromagnetic images of non-cooperative targets [1] [20]. In its most simple form, ISAR systems produce 2D images of targets. A 2D-ISAR image can be interpreted as a filtered projection of a 3D target's reflectivity function onto the image plane. Given the dependence of the image plane orientation on the radar-target geometry and dynamics (which are typically unknown), such projection cannot be predicted. This often results in a difficult interpretation of an ISAR image. The lack of knowledge of the projection of the target onto the image plane necessarily causes difficulties in the classification or recognition of targets by using ISAR images. Although there have been some attempts to estimate the orientation of the image projection plane [15], which directly relates to the estimation of the effective rotation vector, the applicability and the effectiveness of such techniques are yet to be proven. A radical solution to this problem is to form 3D ISAR images, which completely eliminates the problem of dealing with an unknown projection. Previous attempts to form 3D ISAR images can be found in the literature. One of such approaches aims at forming 3D ISAR images by exploiting single sensor ISAR image sequences [6] [17]. 3D target motions, in fact, produce a set of view angles that allow for the estimation of the 3D position of each target scattering center. This approach has the advantage of requiring a single sensor, although it relies on long target observation time intervals and on the a priori knowledge of the target's motions [14]. Another approach is based on interferometric principles and makes use of multiple sensors [19] [2] [8] [21]. Such an approach has the advantage of not requiring long observation time intervals nor the a priori knowledge of the target motions. Classic interferometric techniques uses range profiles and are based on the assumption that a single scatterer is present in a resolution cell. Therefore, the probability of distinguishing more than one scatterer in a range resolution cell is much lower than in the case of a 2D (range and Doppler) resolution cell, obtained when using 2D ISAR imaging. The layover effects are then minimized due to the higher probability of discriminating more than one scatterer in a single range resolution cell. As a consequence, the estimation of a scatterer's position along the cross-range might be improved.

The problem of 3D target reconstruction is addressed and solved in this chapter by exploiting the estimation of the target's effective rotation vector (modulus and phase), which allows the ISAR image plane to be estimated. A partial theoretical foundation of this approach has been introduced in [19] [2]. Nevertheless, some critical aspects still remain either unchallenged or not well defined, such as the scattering center extraction and the 3D reconstruction alignment. Paying particular attention to such problems, a solid theoretical foundation for 3D-interferometric ISAR imaging is provided in this chapter. The scattering center extraction is a fundamental step as it allows the computational burden to be greatly reduced. In fact, by extracting a number of bright scatterers from an ISAR image, the target information can be significantly compressed. As a consequence, the 3D reconstruction reduces into the localization of bright scattering points in a 3D space via the estimation of each scatterer height. The scattering center extraction is carried out in this work by applying an extended version of the CLEAN technique [18], namely the multichannel CLEAN technique (MC-CLEAN). Once the target 3D reconstruction is accomplished, it is important to align the 3D cloud of points with respect to a preset reference system. This operation is fundamental as it allows reconstructed targets to be compared directly with the target's models. In this chapter, a 3D target alignment process that is based on the estimation of the effective rotation vector is detailed. Simulated data are finally used to show some results. The simulated data are obtained by generating the backscattered signal from two point-like scatterer models: an airplane and a boat. By varying the target's motion and the system geometry, several scenarios can be produced. In order to take into account real scenarios, the cases of squinted target and non-orthogonal baselines are introduced.

The organization of this chapter is as follows: the received signal model is discussed in Section 6.1. The steps to be performed in order to obtain a 3D image of the object are detailed in Section 6.2. This includes the multi-channel autofocus technique, the multi-channel CLEAN technique and the 3D reconstruction algorithm. Section 6.3 and Section 6.4 describe the performance analysis and simulation results, respectively.

6.1 MULTI-CHANNEL ISAR SIGNAL MODEL

This section is devoted to presenting the geometry of an interferometric ISAR system and to introducing the multi-channel signal model.

6.1.1 SYSTEM GEOMETRY

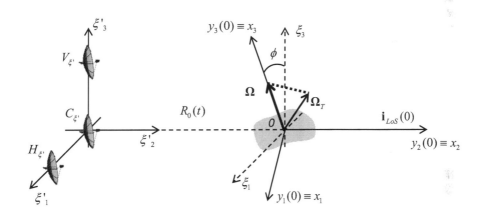

Figure 6.1 ISAR system geometry

The system described in Figure 6.1 consists of three antennas arranged in an L-shape configuration with vertical baseline d_V and horizontal baseline d_H. Having to deal with three reference systems, we will denote points in the 3D space by using a subscript according to the specific reference system coordinates, e.g., $\mathbf{C}_\xi$ and $\mathbf{C}_y$ denote the same point expressed with the reference system coordinates T_ξ and T_y, respectively. The point $\mathbf{C}_{\xi'}$ corresponds to the origin of the reference system T'_ξ, which is the antenna array phase center. The antennas located at points $\mathbf{V}_{\xi'}$ and $\mathbf{H}_{\xi'}$ lie on the ξ'_3 and ξ'_1 axis, respectively. As described in Chapter 2, the reference system T'_ξ is embedded on the radar system while T_y is the time-varying reference system embedded on the target and T_x is coincident with T_y at time $t = 0$. The reference system T_ξ corresponds to T_ξ', but it is centered on the target. For simplicity of representation, let us consider the three antennas to be both transmitting and receiving. Such a configuration would require the use of orthogonal waveforms in order to separate the three channels. However, equal effectiveness could be achieved by considering the use of one transmitter and a certain number of co-located receivers.

Finally, the target is described as a rigid body fixed with respect to T_x, composed of K point-like scatterers with complex amplitude σ_k and with position $\mathbf{x} = [x_{1_k}, x_{2_k}, x_{3_k}]$ at time $t = 0$ for the generic k^{th} scatterer.

6.1.2 RECEIVED SIGNAL MODELING

As described in Chapter 3, the signal received by the ith channel, in the frequency/slow time domain at the output of the matched filter after the application of the *straight iso − range* approximation, can be expressed as

$$S^{(i)}(f, n) = C \cdot W(f, n) \int_V f'(\mathbf{x}) \exp\left\{-j\frac{4\pi f}{c}\left[R_0^{(i)}(n) + \left(\mathbf{x} \cdot \mathbf{i}_{LOS_x}^{(i)}(n)\right)\right]\right\} d\mathbf{x}$$

(6.1)

where

$$W(f, n) = \text{rect}\left[\frac{n}{N}\right] \text{rect}\left[\frac{f - f_0}{B}\right]$$

(6.2)

defines the domain where the two-dimensional Fourier transform (2D-FT) of the reflectivity function $f'(\mathbf{x})$ is defined, C is the complex amplitude, f_0 is the carrier frequency, B is the transmitted signal bandwidth, N is the number of pulses transmitted by the radar, $R_0^{(i)}(n)$ is the modulus of vector $\mathbf{R}_0^{(i)}(n)$ which locates the position of the focusing point O, $\mathbf{i}_{LOS_x}^{(i)}(n)$ is the LoS unit vector of $\mathbf{R}_0(n)$ expressed with respect to T_x and $i \in \{V, C, H\}$. The function $\text{rect}(x)$ is equal to 1 for $|x| < 0.5$, 0 otherwise.

For an ideal point scatterer, Equation (6.1) is modified as follows:

$$S^{(i)}(f, n) = \sigma_k^{(i)} \exp\left\{-j\frac{4\pi f}{c}R_0^{(i)}(n)\right\} \exp\left\{-j\frac{4\pi f}{c}\left[\mathbf{x} \cdot \mathbf{i}_{LOS_x}^{(i)}(n)\right]\right\} W(f, n)$$

(6.3)

When the baselines are short compared to the radar-target distance, the reflectivity functions $\sigma_k^{(i)}$ can be considered the same for all the channels, i.e., $\sigma_k^{(i)} = \sigma_k$.

Since the scalar product $\mathbf{y}(n) \cdot \mathbf{i}_{LOS_y}^{(i)}(n)$ is invariant with respect to the chosen reference system, the scalar product in Equation (6.3) can be rewritten in the T_y reference system. Then, the received signal after motion compensation can be written as follows:

$$S_C^{(i)}(f,n) = \sigma_k \exp\left\{-j\frac{4\pi f}{c}\left[\mathbf{y}(n)\cdot\mathbf{i}_{LOS_y}^{(i)}(n)\right]\right\}W(f,n) \tag{6.4}$$

where $\mathbf{y}(n)$ is the position of an arbitrary target's point scatterer and $\mathbf{i}_{LOS_y}^{(i)}(n)$ is the LoS unit vector expressed with respect to T_y. The distance between the focusing point O and the projection of the scatterer onto the LoS is mathematically expressed by the scalar product $\mathbf{y}(n)\cdot\mathbf{i}_{LOS_y}^{(i)}(n)$.

Under the assumption of a small observation time, the total rotation vector can be considered as constant and the image plane fixed with respect to T_ξ:

$$\mathbf{\Omega}_T(n) \cong \mathbf{\Omega}_T, \quad 0 \le nT_R \le T_{obs}. \tag{6.5}$$

In this case, the position of the target's point scatterer $\mathbf{y}(n)$ can be calculated by solving the following differential equation system:

$$\begin{cases} \dot{\mathbf{y}}(n) = \mathbf{\Omega}_T \times \mathbf{y}(n) \\ \mathbf{y}(0) = \mathbf{x} \end{cases} \tag{6.6}$$

It is worth noting that the position of the scatterer in Equation (6.6) is referred to as the T_y reference system.

The resulting closed form solution is as follows [3, 16, 2]:

$$\mathbf{y}(n) = \mathbf{a} + \mathbf{b}\cos(\Omega_T nT_R) + \frac{\mathbf{c}}{\Omega_T}\sin(\Omega nT_R) \tag{6.7}$$

where $\mathbf{a} = \frac{(\mathbf{\Omega}_T\cdot\mathbf{x})}{\Omega^2}\mathbf{\Omega}_T$, $\mathbf{b} = \mathbf{x} - \frac{(\mathbf{\Omega}_T\cdot\mathbf{x})}{\Omega^2}\mathbf{\Omega}_T$, $\mathbf{c} = \mathbf{\Omega}_T \times \mathbf{x}$ and $\Omega_T = |\mathbf{\Omega}_T|$ and $\mathbf{\Omega}_T = (0, \Omega_{T_2}, \Omega)$ in the chosen reference systems. Since Ω_{T_2} is aligned along the radar LoS, it does not produce any aspect angle variation and therefore does not contribute to the ISAR image formation. Under the usually verified hypothesis of short observation time and small aspect angle variation, Ω_{T_2} is usually much smaller than Ω and as a consequence Equation (6.7) can be rewritten as follows:

$$\mathbf{y}(n) = \mathbf{a} + \mathbf{b}\cos(\Omega nT_R) + \frac{\mathbf{c}}{\Omega}\sin(\Omega nT_R) \tag{6.8}$$

The assumption of a small T_{obs} allows also to approximate the term $\mathbf{y}(n)$ in Equation (6.7) by its first-order Taylor series around $t = 0$. The result is expressed in Equation (6.9):

$$\mathbf{y}(n) \cong \mathbf{a} + \mathbf{b} + \mathbf{c}nT_R = \mathbf{x} + \mathbf{c}nT_R \tag{6.9}$$

Under this approximation the ISAR image can be successfully reconstructed by the range-Doppler (RD) technique [20] [11] [4] because the Doppler frequency of each scatterer is considered constant. When the effective rotation vector is sufficiently constant and when the total aspect angle variation is sufficiently small, the RD technique provides focused images. Consequently, a 2D-FT can be used to form the ISAR image. The matrix $M_{\xi y}$ in Equation (6.10) describes the rotation of the reference system T_y of an angle ϕ with respect to T_ξ:

$$
M_{\xi y} = \begin{bmatrix} \cos \phi & 0 & \sin \phi \\ 0 & 1 & 0 \\ -\sin \phi & 0 & \cos \phi \end{bmatrix}
\tag{6.10}
$$

By means of the rotation matrix $M_{\xi y}$, the LoS unit vectors $i^{(i)}_{LOS_y}(n)$ can be written as the normalized difference between the positions of each sensor and the origin of T_x:

$$
i^V_{LOS_y}(n) \triangleq \frac{\mathbf{y}(n) - \mathbf{V}_y(n)}{|\mathbf{y}(n) - \mathbf{V}_y(n)|} = \begin{bmatrix} \frac{-d_V \sin \phi}{\sqrt{R_0(n)^2 + d_V^2}} & \frac{R_0(n)}{\sqrt{R_0(n)^2 + d_V^2}} & \frac{-d_V \cos \phi}{\sqrt{R_0(n)^2 + d_V^2}} \end{bmatrix}
$$

$$
i^C_{LOS_y}(n) \triangleq \frac{\mathbf{y}(n) - \mathbf{C}_y(n)}{|\mathbf{y}(n) - \mathbf{C}_y(n)|} = \begin{bmatrix} 0 & 1 & 0 \end{bmatrix}
\tag{6.11}
$$

$$
i^H_{LOS_y}(n) \triangleq \frac{\mathbf{y}(n) - \mathbf{H}_y(n)}{|\mathbf{y}(n) - \mathbf{H}_y(n)|} = \begin{bmatrix} \frac{-d_H \cos \phi}{\sqrt{R_0(n)^2 + d_H^2}} & \frac{R_0(n)}{\sqrt{R_0(n)^2 + d_H^2}} & \frac{d_H \sin \phi}{\sqrt{R_0(n)^2 + d_H^2}} \end{bmatrix}
$$

where $\mathbf{V}_y(n)$, $\mathbf{C}_y(n)$ and $\mathbf{H}_y(n)$ are the positions of the antennas with respect to T_y and d_V and d_H represent the vertical and horizontal baseline lengths, respectively. Therefore, the scalar product in Equation (6.4) can be written as [4]:

$$
\mathbf{y}(n) \cdot i^V_{LOS_y}(n) \cong (\mathbf{x} + \mathbf{c}nT_R) \cdot i^V_{LOS_y}(n) =
$$

$$
= (x_1 + c_1 nT_R) \left(\frac{-d_V \sin \phi}{\sqrt{R_0(n)^2 + d_V^2}} \right) + (x_2 + c_2 t) \left(\frac{R_0(n)}{\sqrt{R_0(n)^2 + d_V^2}} \right) +
$$

$$
(x_3 + c_3 nT_R) \left(\frac{-d_V \cos \phi}{\sqrt{R_0(n)^2 + d_V^2}} \right) \cong
\tag{6.12}
$$

$$
\cong x_2 + c_2 nT_R - \frac{d_V}{R_0} [(x_1 + c_1 nT_R) \sin \phi + (x_3 + c_3 nT_R) \cos \phi] =
$$

$$
K_0^V + K_1^V nT_R
$$

where

$$
K_0^V \triangleq x_2 - \frac{d_V}{R_0} (x_1 \sin \phi + x_3 \cos \phi)
\tag{6.13}
$$

$$
K_1^V \triangleq c_2 - \frac{d_V}{R_0} (c_1 \sin \phi + c_3 \cos \phi)
\tag{6.14}
$$

The term $R_0(t)$ can be reasonably approximated with $R_0(0) = R_0$ when considering a small observation time. Equivalently, the denominators $\sqrt{R_0(n)^2 + d_V^2}$ and $\sqrt{R_0(n)^2 + d_H^2}$ are approximated as R_0. The terms c_1, c_2 and c_3 are the three components of the vector $\mathbf{c}$ introduced in Equation (6.7). The scalar products for the other two elements can be similarly computed as follows:

$$
\mathbf{y}(n) \cdot i^C_{LOS_y}(n) \cong K_0^C + K_1^C nT_R
\tag{6.15}
$$

$$
\mathbf{y}(n) \cdot i^H_{LOS_y}(n) \cong K_0^H + K_1^H nT_R
\tag{6.16}
$$

where

$$K_0^C \triangleq x_2 \tag{6.17}$$

$$K_1^C \triangleq c_2 \tag{6.18}$$

$$K_0^H \triangleq x_2 + \frac{d_H}{R_0}(x_3 \sin\phi - x_1 \cos\phi) \tag{6.19}$$

$$K_1^H \triangleq c_2 + \frac{d_H}{R_0}(c_3 \sin\phi - c_1 \cos\phi) \tag{6.20}$$

By substituting the scalar products of Equations (6.12), (6.15) and (6.16) into Equation (6.4), the received signal model can be written as follows:

$$S_C^{(i)}(f,n) = \sigma_k \exp\left\{-j\frac{4\pi f}{c}\left[K_0^{(i)} + K_1^{(i)} nT_R\right]\right\} W(f,n) =$$
$$= \sigma_k \exp\left\{-j\frac{4\pi f K_1^{(i)}}{c} nT_R\right\} \mathrm{rect}\left(\frac{n}{N}\right) \exp\left\{-j\frac{4\pi K_0^{(i)}}{c} f\right\} \mathrm{rect}\left(\frac{f-f_0}{B}\right) \tag{6.21}$$

In the case of small aspect angle variations, the standard procedure for obtaining ISAR images is the range-Doppler technique. A 2D-FT is applied and the analytical form of the complex ISAR image in the delay-time (τ) and Doppler (ν) domain is obtained. The result is shown in Equation (6.22).

$$I^{(i)}(\tau,\nu) = RD_{\substack{f\to\tau\\n\to\nu}}\left\{S_C^{(i)}(f,n)\right\} = BNT_R\sigma_k e^{j2\pi f_0\left(\tau-\frac{2}{c}K_0^{(i)}\right)}$$
$$\mathrm{sinc}\left[T_{obs}\left(\nu+\frac{2f_0}{c}K_1^{(i)}\right)\right]\mathrm{sinc}\left[B\left(\tau-\frac{2}{c}K_0^{(i)}\right)\right] \tag{6.22}$$

where $\mathrm{sinc}(x) \triangleq \frac{\sin(\pi x)}{\pi x}$ and $RD_{\substack{f\to\tau\\n\to\nu}}\{\cdot\}$ indicates the operation of image formation by means of the range-Doppler approach.

6.2 3D INISAR IMAGE FORMATION CHAIN

The block diagram given in Figure 6.2 shows the block diagram of the 3D reconstruction processing when using the interferometric method. The signal received from the three antennas is compensated by using the multi-channel image contrast based algorithm (M-ICBA). The scattering centers are then extracted from the ISAR images by means of the multi-channel CLEAN (MC-CLEAN) technique. The interferometric phases measured from the two orthogonal baselines are used to jointly estimate the magnitude of the target's effective rotation vector and the heights of the scattering centers with respect to the image plane. Finally, a 3D image of the moving target is reconstructed from the 3D spatial coordinates of the scattering centers. All the above-mentioned steps will be described in detail in the following sections.

6.2.1 MULTI-CHANNEL AUTOFOCUSING TECHNIQUE

To form effective ISAR images of a moving target, autofocusing techniques must be applied first. The autofocusing algorithm consists of removing a phase

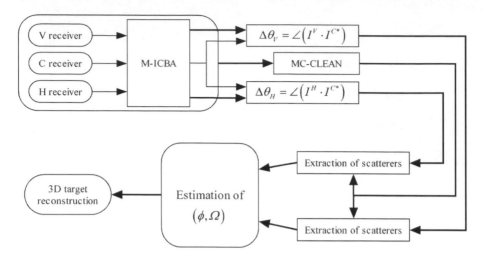

Figure 6.2 Overall flowchart of 3D InISAR reconstruction processing

term from the received signal produced by the target radial motion. Such a phase term is responsible for range migration and Doppler spread, which cause distortions and prevent a proper interpretation of the ISAR image. After motion compensation, the ISAR image of each receiving channel is formed by means of the range-Doppler (RD) algorithm. To effectively apply the 3D reconstruction algorithm, a multistatic autofocus algorithm should be used, which implies that the ISAR images from different spatial channels are all focused with respect to the same focusing point on the target.

Under usually verified hypotheses (target distance much greater than the baselines and receiving antennas lying on a plane orthogonal to the LoS) the phase term responsible for the image distortions can, however, be considered the same for each receiving channel. In these conditions, standard motion compensation algorithms can be applied to an ISAR image [10] and then the estimated phase term used to form the ISAR images relative to the other receiving antennas.

Otherwise, other multistatic autofocusing algorithms, like that described in [5], must be applied.

The autofocus technique adopted here is the image contrast based autofocus (ICBA) algorithm. Such a technique is implemented in two steps: (i) preliminary estimation of the focusing parameters, which are provided by an initialization technique that makes use of the Radon transform (RT) and of a semi-exhaustive search; and (ii) fine estimation, which is obtained by solving an optimization problem where the function to be maximized is the image contrast (IC).

The IC is defined for the receiver in $\mathbf{C}_{\xi'}$ as follows:

$$IC(v_r, a_r) = \frac{\sqrt{A_v\left\{[I^C(\tau, \nu; v_r, a_r) - A_v\{I^C(\tau, \nu; v_r, a_r)\}]^2\right\}}}{A_v\{I^C(\tau, \nu; v_r, a_r)\}} \quad (6.23)$$

where I^C is the intensity of the ISAR image at the receiver located in $\mathbf{C}_{\xi'}$ after the RD gating over the time delay (τ) and Doppler (ν) coordinates, the operator $A_v[\cdot]$ is the average operator, v_r and a_r represent the focusing parameters corresponding to the radial speed and acceleration.

The function IC represents the normalized effective power of the image intensity and gives a measure of the image focusing. In fact, when the image is focused correctly, it is composed of several pronounced peaks (one for each scatterer), which enhance the contrast. When the image is defocused, the image intensity levels are concentrated around the mean value and the contrast is low.

Therefore, the following optimization problem must be solved:

$$(\hat{v}_r, \hat{a}_r) = arg\left(\max_{v_r, a_r}[IC(v_r, a_r)]\right) \quad (6.24)$$

The same focusing parameters can be applied to the other ISAR images formed from the receivers located $\mathbf{V}_{\xi'}$ and $\mathbf{H}_{\xi'}$ receivers. This is because the estimates of the motion parameters can be assumed the same for all the receivers.

Further details about ICBA technique can be found in Section 3.2.3.1.

6.2.1.1 Phase Compensation for Squinted Geometry

The 3D reconstruction of a target in a squinted configuration can be derived by processing the received signal in order to lead it back to the non-squinted configuration.

The target is in a squinted configuration when the line between the origin of the radar reference system and the target is not orthogonal to the plane containing the antennas as shown in Figure 6.3.

If the geometry of the system as well as the position of the target are completely known, we can deduce the equation of the plane perpendicular to the LoS of the center channel. The equation of that plane is described as follows:

$$P_1\xi'_1 + P_2\xi'_2 + P_3\xi'_3 = 0 \quad (6.25)$$

where P_1, P_2 and P_3 are the coordinates of the target defined on the reference system $T_{\xi'} = [\xi'_1, \xi'_2, \xi'_3]$.

Consequently, the position of the generic receiving channel can be projected along the respective LoS and the intersection with the plane can be found. Those intersections represent the positions of the equivalent receivers.

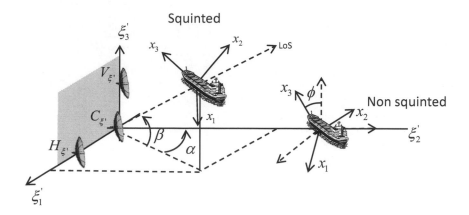

Figure 6.3   Acquisition geometry: squinted and non-squinted configuration

It is worth pointing out that if the target is squinted only on a plane (horizontal or vertical), the equivalent channel defines an effective baseline that is still orthogonal with respect to the other baseline. However, this is no longer true when the target is squinted on both planes as shown in Figure 6.4.

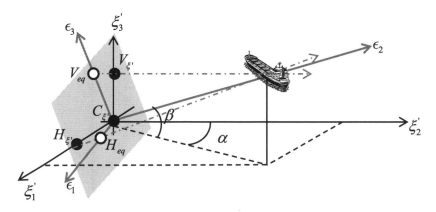

Figure 6.4   Effective receivers geometry

Let the straight line going through the points $P = [P_1, P_2, P_3]$ and $H = [d_H, 0, 0]$ be be written in Cartesian form as follows:

$$\begin{cases} P_2\xi_1' - (P_1 - d_H)\xi_2' - P_2 d_H = 0 \\ -P_3\xi_2' + P_2\xi_3' = 0 \end{cases} \tag{6.26}$$

Then, the position of the horizontal equivalent channel H_{eq} can be found by solving the intersection between the plane in Equation (6.25) and the line in Equation (6.27) as follows:

$$
\begin{bmatrix}
P_2 & -P_1 + d_H & 0 \\
0 & -P_3 & P_2 \\
P_1 & P_2 & P_3
\end{bmatrix}
\cdot
\begin{bmatrix}
\xi_1' \\
\xi_2' \\
\xi_3'
\end{bmatrix}
=
\begin{bmatrix}
P_2 d_H \\
0 \\
0
\end{bmatrix}
\tag{6.27}
$$

The result can be expressed as:

$$
\begin{aligned}
\xi_1' &= \frac{d_H\left(P_2^2 + P_3^2\right)}{P_1^2 + P_2^2 + P_3^2 - d_H P_1} \\
\xi_2' &= -\frac{d_H P_1 P_2}{P_1^2 + P_2^2 + P_3^2 - d_H P_1} \\
\xi_3' &= -\frac{d_H P_1 P_3}{P_1^2 + P_2^2 + P_3^2 - d_H P_1}
\end{aligned}
\tag{6.28}
$$

Let a new reference system $T_\epsilon = [\epsilon_1, \epsilon_2, \epsilon_3]$ be defined as $T_{\xi'}$ after a rotation so that the LoS is coincident to the axis ξ_2'. The matrix describing this rotation can be expressed as composition of the rotation matrices along ξ_1' and ξ_3' as described in Equation (6.29):

$$
M_{VH} = M_V \cdot M_H =
\begin{bmatrix}
\cos\alpha & -\sin\alpha & 0 \\
\cos\beta\sin\alpha & \cos\beta\cos\alpha & \sin\beta \\
-\sin\alpha\sin\beta & -\cos\alpha\sin\beta & \cos\beta
\end{bmatrix}
\tag{6.29}
$$

where

$$
M_V =
\begin{bmatrix}
1 & 0 & 0 \\
0 & \cos\beta & \sin\beta \\
0 & -\sin\beta & \cos\beta
\end{bmatrix}
\tag{6.30}
$$

$$
M_H =
\begin{bmatrix}
\cos\alpha & -\sin\alpha & 0 \\
\sin\alpha & \cos\alpha & 0 \\
0 & 0 & 1
\end{bmatrix}
\tag{6.31}
$$

where α and β are the horizontal and vertical squint angles as shown in Figure 6.3. Let H_ϵ be defined as the horizontal equivalent channel H_{eq} expressed with respect to the new reference system T_ϵ. In order to express the coordinates of H_ϵ, the elements of the rotation matrix M_{VH} must be written with respect to $T_{\xi'}$ as follows:

$$
\begin{aligned}
\cos\alpha &= \frac{P_2}{\sqrt{P_1^2 + P_2^2}} \\
\sin\alpha &= \frac{P_1}{\sqrt{P_1^2 + P_2^2}} \\
\cos\beta &= \frac{\sqrt{P_1^2 + P_2^2}}{\sqrt{P_1^2 + P_2^2 + P_3^2}} \\
\sin\beta &= \frac{P_3}{\sqrt{P_1^2 + P_2^2 + P_3^2}}
\end{aligned}
\tag{6.32}
$$

Thus, by substituting Equation (6.32) into Equation (6.29), the coordinates of H_ϵ can be derived as follows:

$$H_\epsilon = M_{VH} \cdot H_{eq} = \begin{bmatrix} \dfrac{d_H P_2 (P_1^2 + P_2^2 + P_3^2)}{\sqrt{P_1^2 + P_2^2}(P_1^2 - d_H P_1 + P_2^2 + P_3^2)} \\ 0 \\ -\dfrac{d_H P_1 P_3 \sqrt{P_1^2 + P_2^2 + P_3^2}}{\sqrt{P_1^2 + P_2^2}(P_1^2 - d_H P_1 + P_2^2 + P_3^2)} \end{bmatrix} \tag{6.33}$$

The same calculus can be carried out for the vertical channel, but the steps are neglected here. The result is shown in Equation (6.34):

$$V_\epsilon = M_{VH} \cdot V_{eq} = \begin{bmatrix} 0 \\ 0 \\ \dfrac{d_V \sqrt{P_1^2 + P_2^2}\sqrt{P_1^2 + P_2^2 + P_3^2}}{P_1^2 + P_2^2 + P_3^2 - d_V P_3} \end{bmatrix} \tag{6.34}$$

It is worth noting that the coordinates of the equivalent channels in Equations (6.33) and (6.34), in a squinted geometry, represent the position of the receivers after their projection onto the plane which makes the geometry non-squinted. Furthermore, the coordinates of the equivalent channels are not orthogonal anymore.

Within the reference system T_ϵ, the effective baselines can be calculated as follows:

$$\begin{aligned} d_H^{eff} &= \|H_\epsilon - C_\epsilon\| \\ d_V^{eff} &= \|V_\epsilon - C_\epsilon\| \end{aligned} \tag{6.35}$$

Assuming the geometry is completely known, the distance $\Delta^{(i)}(n)$ between the ith equivalent channel and its projection along its LoS onto the plane (which makes the LoS orthogonal to the plane itself) can be easily computed. Therefore, the following phase term can be obtained:

$$H(f, n) = e^{\pm j \frac{2\pi f}{c} \cdot 2\Delta^{(i)}(n)} \tag{6.36}$$

where the sign is negative in the case that the effective channel is farthest from the target than the actual one.

In order to compensate the phase term due to $\Delta^{(i)}(n)$, the matrix $H(f, n)$ is multiplied by the received signal as follows:

$$S'_C(f, n) = S_C(f, n) H(f, n) \tag{6.37}$$

Then, in case of a squinted geometry, the signal $S'_C(f, n)$ should be used in place of $S_C(f, n)$.

6.2.2   MULTI-CHANNEL CLEAN TECHNIQUE

The CLEAN technique can be used to extract the target features such as position and complex amplitude of dominant scattering centers. Considering that the system at hand is a multi-channel system, a modified version of the single channel technique [18] is used here. The modification made stems from a similar concept as discussed in [13], where the polarimetric CLEAN (Pol-CLEAN) technique was presented. It is worth pointing out that the novelty of the proposed MC-CLEAN technique lies in its extension to ISAR images obtained from a spatial multi-channel radar configuration.

The overall flowchart is shown in Figure 6.5.

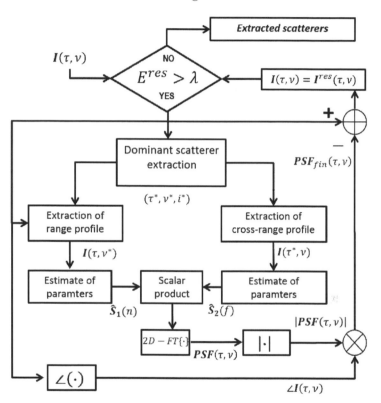

Figure 6.5   Overall flowchart of MC-CLEAN technique

6.2.2.1   Signal Separation

As described in the previous section, the range-Doppler technique is based on an approximation that allows considerations of the support for the received signal in the Fourier domain as a rectangular domain. By means of the RD approximation, the received signal model in Equation (6.21) can be written as the product of two terms, one depending on the slow time index (n) and one depending on the frequency (f), as follows:

$$S_C^{(i)}(f,n) \cong S_1^{(i)}(n)S_2^{(i)}(f) = \sigma_k \exp\left\{-j\frac{4\pi f_0 K_1^{(i)}}{c}nT_R\right\} \text{rect}\left(\frac{n}{N}\right)$$
$$\exp\left\{-j\frac{4\pi K_0^{(i)}}{c}f\right\} \text{rect}\left(\frac{f-f_0}{B}\right) \tag{6.38}$$

where $\text{rect}\left(\frac{n}{N}\right)$ is defined as in Equation (1.17) in Chapter 1. The two components $S_1^{(i)}(n)$ and $S_2^{(i)}(f)$ can be expressed according to the following model:

$$S_1^{(i)}(n) = \sigma_{1_k}^{(i)} \exp\left(j2\pi\left(\eta + f_d nT_R + \frac{\mu}{2}(nT_R)^2\right)\right) \text{rect}\left(\frac{n}{N}\right)$$
$$S_2^{(i)}(f) = \sigma_{2_k}^{(i)} \exp\left(j2\pi f\tau_k\right) \text{rect}\left(\frac{f-f_0}{B}\right) \tag{6.39}$$
$$\sigma_k^{(i)} = \sigma_{1_k}^{(i)}\sigma_{2_k}^{(i)} \cong \sigma_k$$

where f_d is the Doppler frequency, μ is the chirp rate, and τ_k is the time delay associated with the scattering center and η is a constant term which accounts for the constant term distance. It is worth noting that the parameter μ is related to the signal model and accounts for a quadratic radial motion component.

6.2.2.2 Feature Extraction

The MC-CLEAN iteratively:

> Locates the brightest scattering center (dominant scatterer) in one of the multi-channel ISAR images and finds its coordinates in the time delay-Doppler image plane (τ^*, ν^*);
> Removes it from all the ISAR images in order to extract the next dominant scatterer.

In order to eliminate a scattering center from an ISAR image, the scattering center point spread function (PSF) must be estimated and subtracted from the ISAR image.

Let $I^{(i)}(\tau, \nu)$ be the ISAR image reconstructed by the ith receiver.

The dominant scatterer is found within the three images. The range and Doppler indexes, τ^* and ν^*, and the channel i^*, which fomed the image that contains the dominant scatterer, are extracted by means of:

$$(\tau^*, \nu^*, i^*) = \arg\max_{(\tau,\nu,i)}\left\{\left|I^{(i)}(\tau,\nu)\right|\right\} \tag{6.40}$$

with $\tau \in \{1, 2, ..., M\}$, $\nu \in \{1, 2, ..., N\}$ and where M and N are the number of range and Doppler bins.

In order to avoid reselecting the same scatterer to the next iteration, once i^* has been found, we need to estimate its contribution for the other two spatial channels of $I^{(i)}(\tau, \nu)$, so as to delete it from each channel.

By referring to the time component in Equation (6.39), we first estimate the parameters $\sigma_{1_k}^{(i)}$, f_d and μ. The constant η can be neglected because it does not affect the shape of the PSF.

In order to treat the optimization problem in a real domain, the scattering center deletion is performed by using a cost function that depends only on the absolute value of the range profile. Such an operation can be performed in a single channel and then applied to the remaining channels by adjusting the corresponding $\sigma_{1_k}^{(i)}$ parameter. This characteristic is guaranteed by the fact that the PSF is the same for all the ISAR images. It must be pointed out that only the magnitude $\left|\sigma_{1_k}^{(i)}\right|$ of $\sigma_{1_k}^{(i)}$ must be estimated at this stage whereas the phase component is estimated separately and directly in the image domain.

According to [13], the following optimization problem can be stated:

$$\left\{\hat{f}_d, \hat{\mu}, \left|\hat{\sigma}_{1_k}^{i*}\right|\right\} = \arg \min_{\left(f_d, \mu, \left|\sigma_{1_k}^{i*}\right|\right)} \left\{E_{d_{i*}}\left(f_d, \mu, \left|\sigma_{1_k}^{i*}\right|\right)\right\}, \qquad (6.41)$$

where $E_{d_{i*}} = \int |d_{i*}(\nu)|^2\, d\nu$ is the energy of a Doppler slice in the $i*th$ spatial channel, with $d_{i*}(\nu) = \left|I^{(i*)}(\tau^*, \nu)\right| - \left|S_1^{(i*)}(\nu)\right|$ and $S_1^{(i*)}(\nu) = FT_{n \to \nu}\left\{S_1^{(i*)}(n)\right\}$ where the $FT_{n \to \nu}\{\cdot\}$ denotes the discrete Fourier transform.

A similar procedure is followed to estimate the frequency component of the PSF by estimating τ_k and $\sigma_{2_k}^i$ as follows:

$$\left\{\hat{\tau}_k, \left|\hat{\sigma}_{2_k}^{i*}\right|\right\} = \arg \min_{\left(\tau_k, \left|\sigma_{2_k}^{i*}\right|\right)} \left\{E_{g_{i*}}\left(\tau_k, \left|\sigma_{2_k}^{i*}\right|\right)\right\}, \qquad (6.42)$$

where $E_{g_{i*}} = \int |g_{i*}(\tau)|^2\, d\tau$ is the energy of a time-delay section in the $i*th$ spatial channel, with $g_{i*}(\tau) = \left|I^{(i*)}(\tau, \nu^*)\right| - \left|S_2^{(i*)}(\tau)\right|$ and $S_2^{(i*)}(\tau) = IFT_{f \to \tau}\left\{S_2^{(i*)}(f)\right\}$ where the $IFT_{f \to \tau}\{\cdot\}$ denotes the inverse Fourier transform.

The solution of the optimization problem in Equations (6.41) and (6.42) is obtained by using genetic algorithms [12].

The scattering center PSF in the ith spatial channel is obtained by calculating the 2D-FT of the product of the estimates of the time and frequency components multiplied by the phase extracted from the ISAR image, as analytically detailed in:

$$I_{PSF}^{(i)}(\tau, \nu) = \left|2D\!-\!FT_{\substack{f \to \tau \\ t \to \nu}}\left\{\hat{S}_1^{(i)}(n)\hat{S}_2^{(i)}(f)\right\}\right| \angle \left(I^{(i)}(\tau, \nu)\right). \qquad (6.43)$$

Then, at the generic l^{th} iteration, the scattering center must be eliminated from the ISAR image via Equation (6.44) in order to extract the following brightest scatterer:

$$I_{l+1}^{(i)}(\tau, \nu) = I_l^{(i)}(\tau, \nu) - I_{PSF_l}^{(i)}(\tau, \nu). \tag{6.44}$$

The estimation of the PSF is performed by minimizing the image energy after scattering center removal.

The algorithm stops when the residual energy in the ISAR image at the lth iteration is lower than a pre-set threshold Γ. Such a threshold is typically set to a percentage F of the initial energy.

Specifically, the pre-set threshold depends on the energy content and on the SNR of the initial ISAR image, as detailed in Equation (6.45):

$$\Gamma = F \cdot E^{(\mathbf{I}(\tau,\nu))} \frac{SNR}{SNR+1} \tag{6.45}$$

where $E^{(\mathbf{I}(\tau,\nu))} = \sum_i E^{(I^{(i)}(\tau,\nu))}$, with $E^{(I^{(i)}(\tau,\nu))} = \iint |I^{(i)}(\tau, \nu)|^2 d\tau \, d\nu$.

Therefore, the iterations stop when $E_k^{(\mathbf{I}(\tau,\nu))} \frac{SNR}{SNR+1} < \Gamma$.

6.2.3   3D IMAGE RECONSTRUCTION

According to [2], the height of each scatterer with respect to the image plane depends on the angle ϕ and on the phase differences between the antennas along the vertical and horizontal baselines. We also assume that the image projection plane is the same and that the target is located in the far field region.

For the sake of simplicity, in the following sections, both the actual receivers and the effective ones will be written as V, C and H. Similarly, the effective baseline and the actual one will be written as d_H and d_V.

Furthermore, the length of the baselines (or effective baselines in a squinted geometry) is subjected to some constraints. As demonstrated in [23][22], the phase difference between the return echoes received by the orthogonal antennas along either the vertical or the horizontal baselines is a periodic function with period 2π. Thus, the altitude measurement is unambiguous if the baseline lengths satisfy the following upper bounds:

$$d_H \leq \frac{\lambda R_0}{2h_H} \quad , \quad d_V \leq \frac{\lambda R_0}{2h_V} \tag{6.46}$$

where d_H and d_V are the lengths of the baselines, h_H and h_V are the projections of the scatterer with the maximum height onto the imaging plane corresponding to the maximum size of the target, and $\lambda = \frac{c}{f_0}$ is the transmitted wavelength.

It should be pointed out that phase unwrapping is not possible as there is no continuity of phase drift once the bright scatterers are extracted. Therefore, non-ambiguous phase measurements are needed to estimate scatterer's heights.

From the analysis in [23], it is clear that there are conflicting requirements on the baseline length. On one hand, a long baseline is needed in order to minimize phase error measurement. On the other hand, the length of the baseline must be kept short enough in order to minimize the view angle variation.

6.2.3.1 Joint Estimation of the Angle ϕ and Ω

Let the coordinates of the equivalent channels be redefined as follows:

$$\begin{cases} V_\epsilon = [d_V^{\epsilon_1} \ 0 \ d_V^{\epsilon_3}] \\ C_\epsilon = [0 \ 0 \ 0] \\ H_\epsilon = [d_H^{\epsilon_1} \ 0 \ d_H^{\epsilon_3}] \end{cases} \quad (6.47)$$

It should be pointed out that in the case of perfectly orthogonal baselines, the coordinates will result:

$$\begin{cases} V_\epsilon = [0 \ 0 \ d_V^{\epsilon_3}] \\ H_\epsilon = [d_H^{\epsilon_1} \ 0 \ 0] \end{cases} \quad (6.48)$$

The phase differences at the peak of the sinc functions in Equation (6.22) for the horizontal and vertical configurations can be calculated by exploiting the received signals as:

$$\Delta\theta_V = \tfrac{4\pi}{c} f_0 \left(K_0^C - K_0^V \right) =$$
$$= \tfrac{4\pi f_0}{cR_0} [d_V^{\epsilon_1} (x_1 \cos\phi - x_3 \sin\phi) + d_V^{\epsilon_3} (x_1 \sin\phi + x_3 \cos\phi)] \quad (6.49)$$

$$\Delta\theta_H = \tfrac{4\pi}{c} f_0 \left(K_0^C - K_0^H \right) =$$
$$= \tfrac{4\pi f_0}{cR_0} [d_H^{\epsilon_1} (x_1 \cos\phi - x_3 \sin\phi) + d_H^{\epsilon_3} (x_1 \sin\phi + x_3 \cos\phi)] \quad (6.50)$$

The point $\mathbf{x} = \mathbf{y}(0)$ is first mapped onto the point $\boldsymbol{\xi} = [\xi_1, \xi_2, \xi_3]$ from the reference system T_x to the reference system T_ξ. In order to obtain the analytical expression of its coordinates with respect to T_ξ, the following transformation is applied:

$$\boldsymbol{\xi} = \mathbf{M}_{\xi y}^{-1} \cdot \mathbf{y}(0)^T = \mathbf{M}_{\xi y}^{-1} \cdot \mathbf{x} \quad (6.51)$$

where the symbol $(\cdot)^T$ is the transpose operator and the rotation matrix $\mathbf{M}_{\xi y}^{-1}$ is defined as in Equation (6.10).

These coordinates can be expressed as a function of the phase differences. This can be done after rewriting $\Delta\theta_V$ and $\Delta\theta_H$ by substituting Equation (6.51) in Equations (6.49) and (6.50) and then by inverting the obtained equations, as shown below:

$$\begin{cases} \Delta\theta_V = \frac{4\pi f_0}{cR_0} (d_V^{\epsilon_1} \xi_1 + d_V^{\epsilon_3} \xi_3) \\ \Delta\theta_H = \frac{4\pi f_0}{cR_0} (d_H^{\epsilon_1} \xi_1 + d_H^{\epsilon_3} \xi_3) \end{cases} \Rightarrow \begin{cases} \xi_3 = \frac{cR_0}{4\pi f_0} \left(\frac{d_V^{\epsilon_1} \Delta\theta_H - d_H^{\epsilon_1} \Delta\theta_V}{d_V^{\epsilon_3} d_H^{\epsilon_1} - d_V^{\epsilon_1} d_H^{\epsilon_3}} \right) \\ \xi_1 = \frac{cR_0}{4\pi f_0} \left(\frac{d_H^{\epsilon_3} \Delta\theta_V - d_V^{\epsilon_3} \Delta\theta_H}{d_H^{\epsilon_3} d_V^{\epsilon_1} - d_V^{\epsilon_3} d_H^{\epsilon_1}} \right) \end{cases} \quad (6.52)$$

where $\xi_1 = x_1 \cos\phi - x_3 \sin\phi$ and $\xi_3 = x_1 \sin\phi + x_3 \cos\phi$.

Finally, the expression of the coordinates of $\mathbf{x}$ can be obtained by remapping the coordinates from T_ξ to T_x as follows:

$$\mathbf{x} = \mathbf{M}_{\xi \mathbf{y}} \cdot \boldsymbol{\xi} \tag{6.53}$$

In particular, the component x_3 represents the height of the scatterer with respect to the image plane and it can be expressed as a function of the phase differences and the angle ϕ as follows:

$$
\begin{aligned}
x_3 &= \xi_3 \cos\phi - \xi_1 \sin\phi = \\
&\frac{cR_0}{4\pi f_0 \left(d_H^{\epsilon_3} d_V^{\epsilon_1} - d_V^{\epsilon_3} d_H^{\epsilon_1}\right)} \left[(d_V^{\epsilon_1} \triangle\theta_H - d_H^{\epsilon_1} \triangle\theta_V) \cos\phi - (d_H^{\epsilon_3} \triangle\theta_V - d_V^{\epsilon_3} \triangle\theta_H) \sin\phi \right]
\end{aligned}
\tag{6.54}
$$

Furthermore, Ω and ϕ can be jointly estimated by expressing the vector $\mathbf{c}$ in Equation (6.7) with respect to the reference system T_y:

$$\mathbf{c} = \boldsymbol{\Omega}_T \times \mathbf{x} \Rightarrow c_2 \simeq x_1 \Omega \tag{6.55}$$

where $\boldsymbol{\Omega}_T = (0, \Omega_{T2}, \Omega)$, Ω is the modulus of $\boldsymbol{\Omega}$ and Ω_{T2} is the coordinate of $\boldsymbol{\Omega}_T$ along the y_2 axis. It should be noted that this is the result of the selection of T_y. In fact, this reference system is chosen in order to have the y_3 axis aligned with $\boldsymbol{\Omega}$ at $t = 0$, the y_2 axis aligned with the LoS and the y_1 axis to complete a Cartesian triplet. It should also be pointed out that Ω_{T2} does not produce any aspect angle variation, as it is aligned with the LoS. Thus, the first component Ω_{T1} must be zero since Ω is the only component that gives the contribution to changing the aspect angle.

The term c_2 can also be expressed by taking into account the Doppler component as follows:

$$\nu_C \triangleq -\frac{2f_0}{c} K_1^C \cong -\frac{2f_0}{c} c_2 \quad \Rightarrow \quad c_2 = -\nu_C \frac{c}{2f_0} \tag{6.56}$$

By substituting Equation (6.55) into Equation (6.56) and after some algebra we obtain:

$$
\begin{aligned}
\nu_C &= \frac{R_0 \Omega}{2\pi \left(d_H^{\epsilon_3} d_V^{\epsilon_1} - d_V^{\epsilon_3} d_H^{\epsilon_1}\right)} \cdot \\
&\cdot \left[(\triangle\theta_H d_V^{\epsilon_3} - \triangle\theta_V d_H^{\epsilon_3}) \cos\phi + (\triangle\theta_V d_H^{\epsilon_1} - \triangle\theta_H d_V^{\epsilon_1}) \sin\phi \right]
\end{aligned}
\tag{6.57}
$$

Consequently, Equation (6.57) can be rewritten by considering only the contribution due to the kth scatterer as follows:

$$Z_k = aY_k + bX_k \tag{6.58}$$

where $Z \triangleq \frac{\nu_C 2\pi}{R_0}$, $Y \triangleq \frac{\triangle\theta_H d_V^{\epsilon_3} - \triangle\theta_V d_H^{\epsilon_3}}{d_H^{\epsilon_3} d_V^{\epsilon_1} - d_V^{\epsilon_3} d_H^{\epsilon_1}}$, $X \triangleq \frac{\triangle\theta_V d_H^{\epsilon_1} - \triangle\theta_H d_V^{\epsilon_1}}{d_H^{\epsilon_3} d_V^{\epsilon_1} - d_V^{\epsilon_3} d_H^{\epsilon_1}}$, $a \triangleq \Omega \cos\phi$ and $b \triangleq \Omega \sin\phi$.

In other words, Z_k corresponds to the Doppler value of the kth scatterer for the central receiver. Therefore, the interferometric phases are calculated from the matrices $\triangle\theta_H$ and $\triangle\theta_V$ only in the range and Doppler bins where the scatterers are extracted. In this way, the terms X_k, Y_k and Z_k are real values and Equation (6.58) represents the equation of a plane.

Then, the estimates of Ω and ϕ can be calculated by estimating a and b. This can be done by evaluating the regression plane, which is the plane that minimizes the sum of the square distances of the points $P_k = (X_k, Y_k, Z_k)$ from the plane itself.

The problem can be mathematically solved by minimizing the function:

$$\Psi(a, b) = \sum_{k=1}^{K} [Z_k - (aY_k + bX_k)]^2 \qquad (6.59)$$

The estimates $\tilde{a}$ and $\tilde{b}$ of a and b, respectively, are:

$$\left(\tilde{a}, \tilde{b}\right) = \min_{a,b} [\Psi(a, b)] \qquad (6.60)$$

Finally, the estimation of Ω and ϕ can be derived from the estimates $\tilde{a}$ and $\tilde{b}$ as described below:

$$\hat{\Omega} = \sqrt{\tilde{a}^2 + \tilde{b}^2}$$
$$\hat{\phi} = arctan\left(\frac{\tilde{b}}{\tilde{a}}\right) \qquad (6.61)$$

6.3 PERFORMANCE ANALYSIS

It is very important to evaluate the performances of the interferometric processing in order to better interpret the results of the three-dimensional reconstruction.

Consider to analyse simulated data where Gaussian noise has been added to the raw data in order to obtain a given SNR in the data domain.

As described in the previous sections, the inputs of the interferometric processing are the interferometric phases and the range-Doppler positions of the scattering centers extracted by the MC-CLEAN technique. The height of each scatterer with respect to the image plane is then obtained through the estimation of the mentioned parameters. Consequently, the resulting three-dimensional shape is composed of several scattering centers. In order to analyze the performances of this method, the first step is then to realign the reconstructed target, expressed with respect to the image plane, with the reference model, expressed with respect to T_ξ. In this way, each scatterer of the reconstructed target can be consistently compared with each scatterer of the reference model.

The performance analysis block diagram is shown in Figure 6.6.

First, the reference model is rotated so that its heading coincides with the assigned trajectory and the reconstructed target is rotated so that the image plane coincides with the horizontal plane where the reference model is defined.

Second, each scatterer of the 3D reconstructed target is associated to the correspondent scatterer of the model through a soft assignment processing. Finally, performance indicators are defined in order to verify the effectiveness of the method.

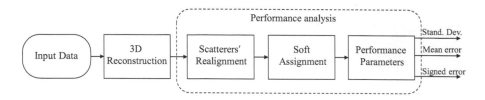

Figure 6.6 Performance analaysis flowchart

6.3.1 SCATTERERS REALIGNMENT

The scatterers realignment is composed of the following two steps:

1. **Rotation of the model along the trajectory**
   The model is rotated along the trajectory by means of the rotation matrix $\boldsymbol{R}_t$ as follows:

$$\boldsymbol{C}_{rt} = \boldsymbol{R}_t \cdot \boldsymbol{C}_m$$
$$\boldsymbol{R}_t = \boldsymbol{R}_\mu \cdot \boldsymbol{R}_\nu \cdot \boldsymbol{R}_\phi \tag{6.62}$$

   where $\boldsymbol{C}_m$ is the matrix of dimensions $3 \times M$ describing the coordinates of the model; $\boldsymbol{R}_\mu, \boldsymbol{R}_\nu$ and $\boldsymbol{R}_\xi$ are the yaw, pitch and roll rotation matrices, respectively. Finally, $\boldsymbol{C}_{rt}$ is the matrix representing the three-dimensional coordinates of the model after being rotated.

2. **Image plane rotation**
   The reconstructed target is rotated in order to have the image plane coincident with the horizontal plane. This can be done by remapping the coordinates of the reconstructed target from the reference system T_x to the reference system T_ξ, as defined in Equation (6.51).

6.3.2 SOFT ASSIGNMENT

The adopted soft assignment method is based on a probabilistic least squares (PLS) approach [9][7]. A soft assignment procedure associates each model scatterers to each reconstructed scatterers. Each of these assignments is weighted by a coefficient that is defined as a probability. In fact, the sum of all possible assignments for a given scatterer is 1. Thus, the optimization problem can be expressed as:

$$\hat{\boldsymbol{\alpha}} = \arg\{\min\{J(\boldsymbol{\alpha})\}\} \tag{6.63}$$

where

$$J\left(\boldsymbol{\alpha}\right) = \sum_{i=1}^{K_M}\sum_{k=1}^{K}\alpha_{i,k}\varepsilon_{i,k}^{T}\varepsilon_{i,k}\alpha_{i,k} \qquad (6.64)$$

is the cost function; K is the number of the extracted scattering centers, K_M is the number of scatterers which composes the target model and $\varepsilon_{i,k}$ is the distance between the ith scatterer of the model and the kth scatterer of the reconstructed target; $\boldsymbol{\alpha}$ is the matrix of dimensions $K_M \times K$ containing the soft assignments:

$$\boldsymbol{\alpha} = \begin{bmatrix} \alpha_{1,1} & \cdots & \alpha_{1,K} \\ \vdots & \vdots & \vdots \\ \vdots & \alpha_{i,k} & \vdots \\ \vdots & \vdots & \vdots \\ \alpha_{M,1} & \cdots & \alpha_{M,K} \end{bmatrix} \qquad (6.65)$$

Each element of the matrix $\alpha_{i,k}$ represents the probability that the kth extracted scatterer belongs to the ith model's scatterer and is expressed as follows:

$$\begin{cases} \alpha_{i,k} = \dfrac{\left(\varepsilon_{i,k}^{T}\varepsilon_{i,k}\right)^{-1}}{\sum_{p=1}^{M}\left(\varepsilon_{p,k}^{T}\varepsilon_{p,k}\right)^{-1}}, \\ \sum_{i=1}^{M}\alpha_{i,k} = 1 \quad \forall k = 1,\ldots,K \end{cases} \qquad (6.66)$$

Finally, each extracted scatterer is assigned to the model's scatterer with the highest $\alpha_{i,k}$.

6.3.3 PERFORMANCE INDICATORS

Two different kinds of errors could affect the results: the assignment error and the scatterers' height estimation error.

The assignment error is related to the identification of unreliable assignments and therefore it allows for an accurate evaluation of the scatterers' height estimation error. An unreliable assignment is declared when a scatterer's height error ϵ_h is greater than a fixed threshold Λ, which is expressed as follows:

$$\Lambda = A_v\left[\bar{\epsilon}_h\left(s\right)\right] + \gamma \cdot \sigma_{\epsilon_h} \qquad (6.67)$$

where $A_v[\cdot]$ indicates the expectation operator calculated by considering S Monte Carlo runs of the simulation and $\bar{\epsilon}_h\left(s\right)$ is defined as follows:

$$\bar{\epsilon}_h\left(s\right) = \frac{1}{K}\sum_{k=1}^{K}\epsilon_h\left(k,s\right) \qquad (6.68)$$

where the height error of the single k^{th} scatterer of the sth Monte Carlo step is defined as:

$$\epsilon_h\left(k,s\right) = \left|h_m\left(k,s\right) - h_r\left(k,s\right)\right| \qquad (6.69)$$

where $h_m(k, s)$ and $h_r(k, s)$ are the heights with respect to the image plane referring to the model and to the reconstructed target, respectively. The parameter σ_{ε_h} is the standard deviation of ϵ_h and $\gamma \in \mathbb{R}$ is a parameter that can vary to empirically adjust the threshold Λ. Simulations with different values of γ have demonstrated that this parameter does not affect the final result.

The scatterers' height estimation error is then computed using the same procedure, after having discarded all the unreliable assignments.

Finally, the error on the estimates of the angle $\hat{\phi}$ and the effective rotation vector $\hat{\Omega}$ are calculated by using the expectation operator as follows:

$$A_v[\epsilon_\Omega(s)] = \frac{1}{S} \sum_{s=1}^{S} \epsilon_\Omega(s) \tag{6.70}$$

and

$$A_v[\epsilon_\phi(s)] = \frac{1}{S} \sum_{s=1}^{S} \epsilon_\phi(s) \tag{6.71}$$

where $\epsilon_\Omega(s)$ and $\epsilon_\phi(s)$ are defined as follows:

$$\epsilon_\Omega(s) = \left| \hat{\Omega}(s) - \Omega \right| \tag{6.72}$$

$$\epsilon_\phi(s) = \left| \hat{\phi}(s) - \phi \right| \tag{6.73}$$

where Ω and ϕ are a priori known.

6.4   SIMULATION RESULTS

In this section some results with simulated data are shown aiming at assessing the effectiveness of the proposed 3D InISAR algorithm. Both the squinted and non-squinted configuration have been tested.

6.4.1   SIMULATION SET-UP

The targets are assumed to be a rigid body moving in a rectilinear trajectory (defined by roll, pitch and yaw directions) with respect to the radar LoS.

In both the simulations, the target resembles the shape of a boat and it is composed of $K = 36$ ideal scatterers.

Two simulations are shown.

In the first simulation, perfect motion compensation has been applied. The target is in a non-squinted configuration and the performance analyses are carried out for increasing baseline length and SNR.

In the second simulation, ICBA processing has been applied. The target is in a squinted configuration and the performance analyses are carried out increasing the squint angle and SNR with fixed baselines. In both the cases, target motions are characterized by their own motion components.

The target is shown in Figure 6.7. For a clearer and simpler visualization, its principal axes coincide with the T_x Cartesian coordinates.

The threshold Λ selected for both the simulations has been tested by varying the parameter γ. Results do not show significant variations. A reference value of $\gamma = 1.2$ has been considered.

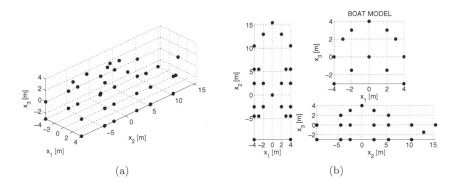

(a)　　　　　　　　　　　　　　　　　(b)

Figure 6.7　Model targets composed of ideal point-like scatterers

Table 6.1 and Table 6.2 show the parameters used in the simulations.

Simulation nr.1

Table 6.1

Parameters of simulation nr.1

Number of freq.	256	γ	1.2
Radar sweeps	128	Target position	$[0, 10, 0] Km$
Bandwidth	300 MHz	Target velocity	10 m/s
Carrier freq.	10 GHz	Roll/Pitch/Yaw	$0°/0°/60°$
T_{obs}	0.8 s	Baselines	$[1, 3, 5] m$

The final result is shown in Figure 6.8 and Figure 6.9, where the reconstructed targets are superimposed onto the models.

Figure 6.10 and Figure 6.11 show the behavior of the height error and the standard deviation of the height error as the SNR and the baseline length change. The height error and the standard deviation of the height error are expressed in meters. The baseline lengths are chosen in order to satisfy the constraint expressed in Equation (6.46). Short baselines produce larger phase error measurements and therefore larger height estimation errors occur. On the other hand, as the baseline length exceeds the upper bounds in Equation

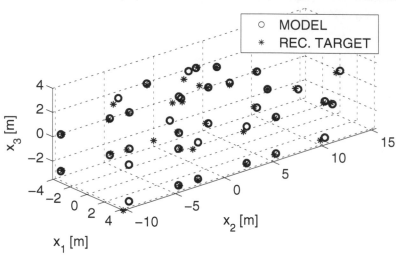

Figure 6.8 Results of the target reconstruction

(6.46), the height error tends to increase since the target is not contained in the unambiguous window.

Figure 6.12 depicts the height error $A_v\left[\bar{\epsilon}_h\left(k\right)\right]$ calculated by preserving the positive or negative sign of ϵ_h, referring to both the considered scenarios. Consistently with the previous results, the values of the error with sign are lower when the unreliable assignments are discarded. Furthermore, it can be noted that the estimator is unbiased.

The same descendant monotonic behavior can be observed in Figure 6.13-6.16, where the mean error and the standard deviation of the estimates of Ω and ϕ are depicted.

It can be observed that the evaluated performance indicators do not follow the described descendant behavior when the baseline length is equal to 5 m. This is because the baseline length exceeds the constraints expressed in Equation (6.46).

Simulation nr.2

It should be noticed that the several target's positions considered correspond to squint angles equal to $[0°, 15°, 30°, 45°, 60°]$ in the horizontal plane.

Figure 6.17 and Figure 6.18 show the behavior of the height error and the standard deviation of the height error as the SNR and the squint angle change. The height error and the standard deviation of the height error are

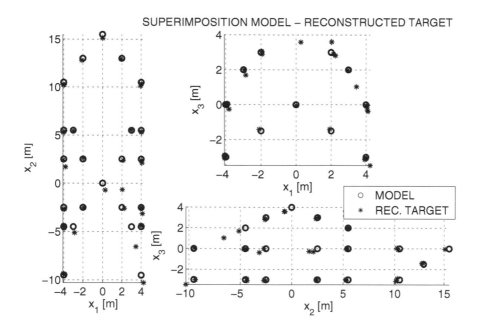

Figure 6.9 Results of the target reconstruction along three different planes

expressed in meters. When the squint angle increases, the effective baseline becomes shorter than the actual one. This is due to the projection of the actual baselines onto the plane orthogonal to the LoS. As expected, short baselines produce larger phase error measurement and therefore larger height estimation errors occur. This is clear in all of the figures.

It is worth pointing out that when the unreliable assignments are discarded, the values of both the height error and the standard deviation decrease significantly for both the simulations. Furthermore, in this case, the curves follow a decreasing trend as the SNR increases and as the baseline length increases, accordingly with the theory. Thus, it can be deducted that the correct identification of the unreliable assignments, and consequently the choice of an appropriate threshold, is fundamental in order to obtain reliable results.

Figure 6.19 depicts the height error $A_v\left[\bar{\epsilon}_h(k)\right]$ calculated by preserving the positive or negative sign of ϵ_h, referring to both the considered scenarios. Consistently with the previous results, the values of the signed error are lower when the unreliable assignments are discarded. Furthermore, it can be noted that the estimator is unbiased, even with further processing due to M-ICBA technique.

The same descendant pattern can be observed in Figure 6.20-6.23, where the mean error and the standard deviation of the estimates of Ω and ϕ are depicted.

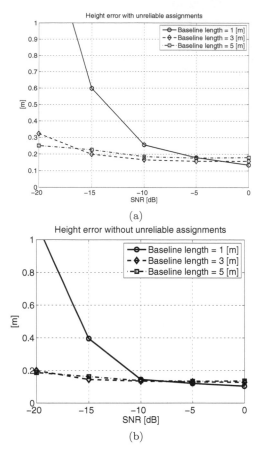

Figure 6.10 Simulation nr.1, height error. (a) Height error with unreliable assignments; (b) height error without unreliable assignments

6.5 CONCLUSIONS

A 3D InISAR image processing method has been discussed in this chapter. First, an interferometric L-shaped system composed of three TX/RX antennas has been presented. For this system, a multi-channel ISAR signal model has been introduced. The main steps to be performed in order to obtain a three-dimensional image of the target have been detailed. In particular, mathematical details regarding the MC-CLEAN and the 3D reconstruction algorithm have been given.

The 3D InISAR algorithm lays on the hypothesis of orthogonality between the LoS and the 2D plane where the transceivers lay on. Such a hypothesis, however, could not always be verified in a real scenario. Then, the squinted

Table 6.2

Parameters of simulation nr.2

Number of freq.	256	γ	1.2
Radar sweeps	128	Target position/Squint angle	$[0, 10, 0]km/0°$ $[2.58, 9.66, 0]km/15°$ $[5, 8.66, 0]km/30°$ $[7.07, 7.07, 0]km/45°$ $[8.66, 5, 0]km/60°$
Bandwidth	300 MHz	Target velocity	10 m/s
Carrier freq.	10 GHz	Roll/Pitch/Yaw	0°/0°/60°
T_{obs}	0.8 s	Baselines	2 m

case is analyzed as well, and a post-processing technique is formulated so as to deal with such a problem.

Furthermore, a performance analysis has been carried out to assess the algorithm performance in both the squinted and non-squinted configuration.

REFERENCES

1. Dale A. Ausherman, Adam Kozma, Jack L. Walker, Harrison M. Jones, and Enrico C. Poggio. Developments in Radar Imaging. *Aerospace and Electronic Systems, IEEE Transactions on*, AES-20(4):363–400, July 1984.
2. N. Battisti and M. Martorella. Intereferometric phase and target motion estimation for accurate 3D reflectivity reconstruction in ISAR systems. In *Radar Conference, 2010 IEEE*, pages 108–112, May 2010.
3. F. Berizzi, E. Dalle Mese, M. Diani, and M. Martorella. High-resolution ISAR imaging of maneuvering targets by means of the range instantaneous Doppler technique: modeling and performance analysis. *Image Processing, IEEE Transactions on*, 10(12):1880–1890, Dec. 2001.
4. F. Berizzi and M. Diani. Target angular motion effects on ISAR imaging. *Radar, Sonar and Navigation, IEE Proceedings*, 144(2):87–95, 1997.
5. Stefan Brisken, Marco Martorella, Torsten Mathy, Christoph Wasserzier, and Elisa Giusti. Multistatic isar autofocussing using image contrast optimization. In *Radar Systems (Radar 2012), IET International Conference on*, pages 1–4, 2012.
6. T. Cooke. Ship 3D model estimation from an ISAR image sequence. In *Radar Conference, 2003. Proceedings of the International*, pages 36–41, Sept. 2003.
7. E. Giusti, M. Martorella, and A. Capria. Polarimetrically-Persistent-Scatterer-Based Automatic Target Recognition. *Geoscience and Remote Sensing, IEEE Transactions on*, 49(11):4588–4599, Nov. 2011.
8. J.A. Given and W.R. Schmidt. Generalized ISAR-part ii: interferometric techniques for three-dimensional location of scatterers. *Image Processing, IEEE Transactions on*, 14(11):1792–1797, Nov. 2005.

9. M.L. Krieg and D.A. Gray. Comparison of probabilistic least squares and probabilistic multi-hypothesis tracking algorithms for multi-sensor tracking. In *Acoustics, Speech, and Signal Processing, 1997. ICASSP-97., 1997 IEEE International Conference on*, volume 1, pages 515–518, Apr. 1997.

10. M. Martorella, F. Berizzi, and B. Haywood. Contrast maximisation based technique for 2-D ISAR autofocusing. *Radar, Sonar and Navigation, IEE Proceedings*, 152(4):253–262, Aug. 2005.

11. Marco Martorella. Introduction to inverse synthetic aperture radar. In *Elsevier Academic Press Library in Signal Processing: Communications and Radar Signal Processing*. (Vol. 2), 1st Ed., Sept. 2013.

12. Marco Martorella, Fabrizio Berizzi, and Silvia Bruscoli. Use of Genetic Algorithms for Contrast and Entropy Optimization in ISAR Autofocusing. *EURASIP Journal on Advances in Signal Processing*, 2006(1):087298, 2006.

13. Marco Martorella, Andrea Cacciamano, Elisa Giusti, Fabrizio Berizzi, Brett Haywood, and Bevan Bates. CLEAN Technique for Polarimetric ISAR. *International Journal of Navigation and Observation*, 2008:1–13, 2008.

14. W. Nel, D. Stanton, and M.Y.A. Gaffar. Detecting 3-D rotational motion and extracting target information from the principal component analysis of scatterer range histories. In *Radar Conference - Surveillance for a Safer World, 2009. RADAR. International*, pages 1–6, Oct. 2009.

15. D. Pastina and C. Spina. Slope-based frame selection and scaling technique for ship ISAR imaging. *Signal Processing, IET*, 2(3):265–276, Sept. 2008.

16. A. Scaglione and S. Barbarossa. Estimating motion parameters using parametric modeling based on time-frequency representations. In *Radar 97 (Conf. Publ. No. 449)*, pages 280–284, Oct. 1997.

17. M. Stuff, M. Biancalana, G. Arnold, and J. Garbarino. Imaging moving objects in 3D from single aperture Synthetic Aperture Radar. In *Radar Conference, 2004. Proceedings of the IEEE*, pages 94–98, Apr. 2004.

18. Y. Sun and P. Lin. An improved method of ISAR image processing. In *Circuits and Systems, 1992., Proceedings of the 35th Midwest Symposium on*, volume 2, pages 983–986, Aug. 1992.

19. Genyuan Wang, Xiang-Gen Xia, and V.C. Chen. Three-dimensional ISAR imaging of maneuvering targets using three receivers. *Image Processing, IEEE Transactions on*, 10(3):436–447, March 2001.

20. D.R. Wehner. *High-Resolution Radar*. Artech House Radar Library. Artech House, 1995.

21. Xiaojian Xu and R.M. Narayanan. Three-dimensional interferometric ISAR imaging for target scattering diagnosis and modeling. *Image Processing, IEEE Transactions on*, 10(7):1094–1102, July 2001.

22. Qun Zhang and Tat Soon Yeo. Three-dimensional SAR imaging of a ground moving target using the InISAR technique. *Geoscience and Remote Sensing, IEEE Transactions on*, 42(9):1818–1828, Sept. 2004.

23. Qun Zhang, Tat Soon Yeo, Gan Du, and Shouhong Zhang. Estimation of three-dimensional motion parameters in interferometric ISAR imaging. *Geoscience and Remote Sensing, IEEE Transactions on*, 42(2):292–300, Feb. 2004.

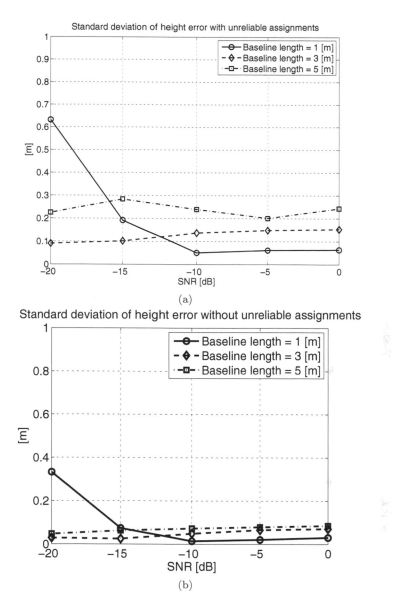

Figure 6.11 Simulation nr.1, standard deviation of the height error. (a) Standard deviation of the height error with unreliable assignments; (b) standard deviation of the height error without unreliable assignments

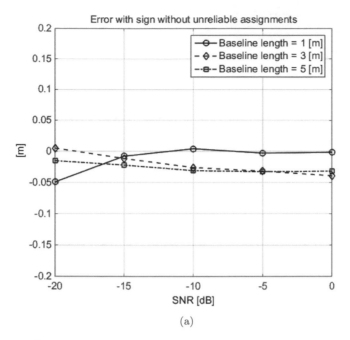

(a)

Figure 6.12 Simulation nr.1, error with sign without unreliable assignments

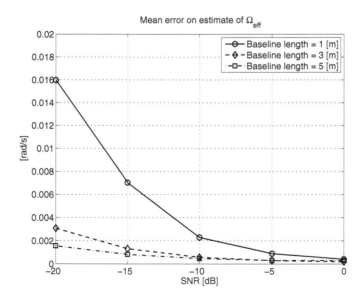

Figure 6.13 Simulation nr.1, mean error of the estimate of Ω

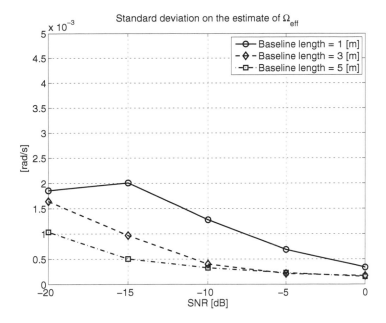

Figure 6.14 Simulation nr.1, standard deviation of the estimate of Ω

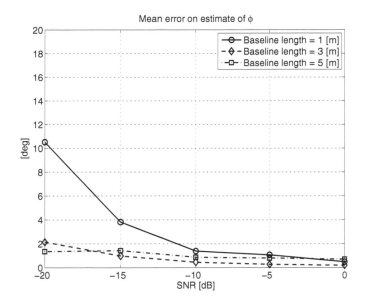

Figure 6.15 Simulation nr.1, mean error of the estimate of ϕ

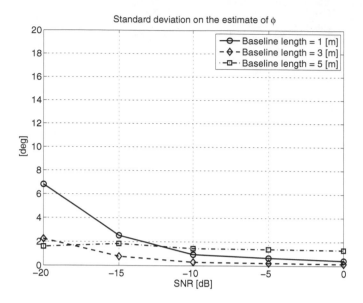

Figure 6.16 Simulation nr.1, standard deviation of the estimate of ϕ

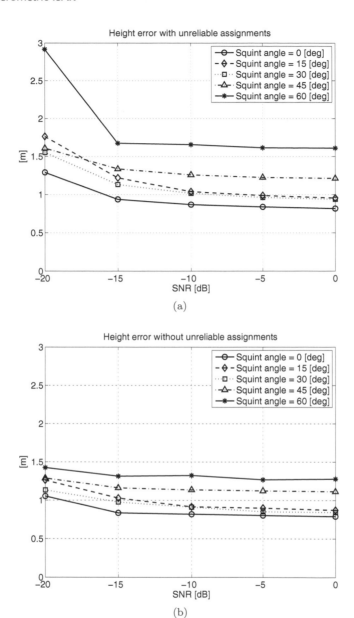

Figure 6.17 Simulation nr.2, height error. (a) Height error with unreliable assignments; (b) height error without unreliable assignments

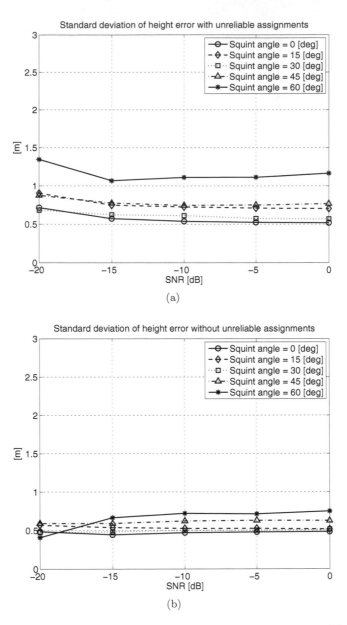

Figure 6.18 Simulation nr.2, standard deviation of the height error. (a) Standard deviation of the height error with unreliable assignments; (b) standard deviation of the height error without unreliable assignments

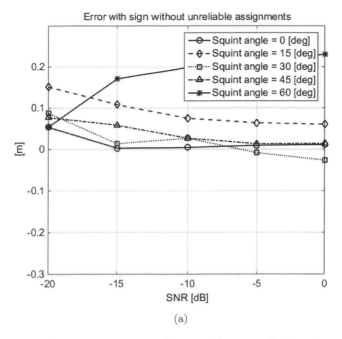

Figure 6.19 Simulation nr.2, error with sign without unreliable assignments

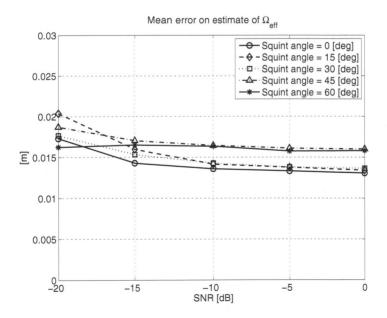

Figure 6.20 Simulation nr.2, mean error of the estimate of Ω

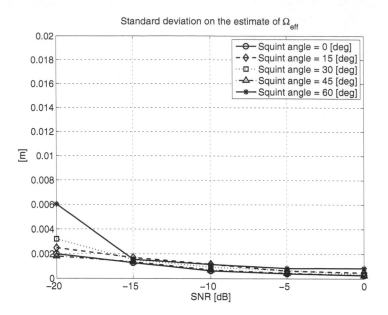

Figure 6.21 Simulation nr.2, standard deviation of the estimate of Ω

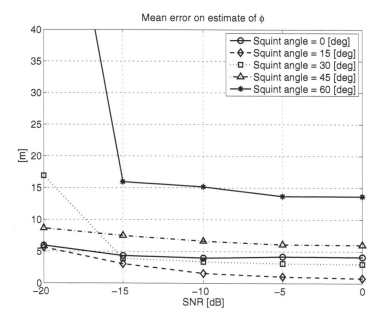

Figure 6.22 Simulation nr.2, mean error of the estimate of ϕ

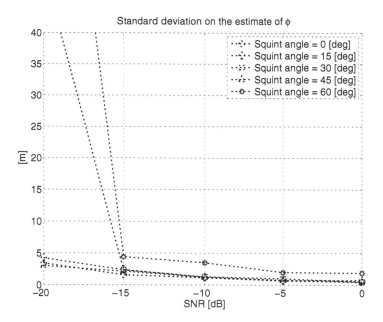

Figure 6.23 Simulation nr.2, standard deviation of the estimate of ϕ

Part II

Applications

7 Detection of Ships from SAR Images

*D. Stagliano, A. Lupidi, M. Martorella, and
F. Berizzi*

CONTENTS

The observation of maritime activity has been a field of research ever since SAR images of the ocean surface became available. This is because SAR images provide global scale coverage, independently of the weather and of the night/day cycle, and high spatial resolutions. More specifically, ship detection and classification from SAR images are effective tools for fishing activity monitoring and detection of ships responsible for marine oil pollution. Furthermore, SAR images are used for detecting illegally operating ships that increase marine crimes including smuggling and sea-jacking by piracy.

However, modern SAR systems generate large amounts of data due to the observation of broad expanses and the high spatial resolutions. Hence, the necessity of an algorithm for automatic ship detection that guarantees both the reliability and the accuracy of detection results. Unfortunately, the presence of speckle noise complicates the automatic interpretation of the SAR images. The detection of maritime targets in coastal areas is even much more complicated as both the bathymetry and the local wind fields deeply affect the sea clutter statistical behaviour. For this reason, the sea clutter may be spatially non-homogeneous. Consequently, its statistical characterization may change within the same SAR image.

Many detection algorithms have been proposed until now [20, 4, 11, 3, 7], some of them based on the CFAR method [13, 14, 5, 15, 1]. Since the detection is not the main focus of this book, one detection technique is only presented, which is based on an adaptive threshold and which has been recently proposed in the scientific literature [16, 17].

Some results obtained by applying this detection technique to spaceborne SAR images are shown in this chapter. The effectiveness of such a technique and the benefits of using global scale SAR images to monitor large areas of interest are proven.

7.1   ALGORITHM DESCRIPTION

The electromagnetic interaction between microwaves and sea surface makes the sea clutter in SAR images with specific statistical properties. However, speckle noise, bathymetry and local winds fields, especially in coastal areas, may complicate the SAR image interpretation. In fact, they deeply affect the homogeneity of the clutter statistics, thus complicating the detection of maritime targets. CFAR processing schemes can be used to overcome this problem. Such methods set the threshold adaptively based on local information of the clutter noise power. The threshold in a CFAR detector is set on a cell by cell basis using estimated clutter power by processing a group of reference cells surrounding the CUT (cell under test). In a homogeneous background when the reference cells contain independent and identically distributed (i.i.d.) observations (governed by exponential distribution) the CA-CFAR processor is the optimum CFAR processor. The CA-CFAR (cell average-CFAR) procedure uses the maximum likelihood estimate of the noise power to adaptively set the threshold under the assumption of i.i.d noise samples. However, the CA-CFAR performances are significantly affected when the assumption of clutter homogeneity is violated. Moreover, conventional CFAR-based algorithms could be very time consuming because they inspect all the pixels of a SAR image.

Then, an algorithm has been proposed in [17] that aims at reducing the computation time. This algorithm can be interpreted as an improved extension of the two-stage algorithm described in [16]. Figure 7.1 shows the block-scheme of the proposed approach, which is composed of three main steps:

- A *pre-processing* algorithm, which aims at improving the image quality to facilitate both the sea/land separation and the maritime targets detection.
- The detector, which is composed of two steps: the *S-detector* and *W-CFAR*.
- A *post-processing* step, which consists of clustering together pixels belonging to the same target. This is an important step since in high resolution SAR images large targets may occupy several resolution cells.

Figure 7.1   Detection algorithm block-scheme

7.1.1   PRE-PROCESSING: WAVELET CORRELATOR

SAR images, especially, maritime SAR images, are typically affected by different kinds of disturbances such as speckle noise and maritime discontinuity effects. The speckle noise is due to cross-scatterers interferences while the maritime discontinuity effects are due to random changes in bathymetry and wind currents. Such disturbances may affect the correct interpretation of the SAR image thus leading to false alarms, missing detections and erroneous target classification and recognition.

All these remarks and the fact that the disturbances may be spatially distributed suggest to study the SAR image spatial statistical behaviour and eventually to use effective despeckle methods prior to apply the detection algorithm.

An interesting observation [18] is that the human eye sees a texture over two fundamental properties: the orientation of its different elements and its frequency content. In other words, it can simultaneously perform a selective filter in frequency and orientation over the observed scene. Moreover, the human visual system perceives images in a multi-scale way, as it is able to focus on different elements at different scales.

As the wavelet transform (WT) provides a multi-scale analysis of images, according to their spatial and frequency characteristics, multi-resolution processing with wavelets represents a suitable tool for modeling the human vision system.

These observations lead us to think that the wavelet theory may be useful to improve the interpretation of a SAR image and then to detect ships.

Differently from clutter and speckle, man-made targets, such as vessels, exhibit a deterministic behavior. This fact explains why the presence of a vessel is usually noticeable in the sub-bands (or scales) obtained when a two-dimensional *discrete wavelet transform* is applied [12].

This kind of discontinuity can even be transmitted over scales. In fact, a vessel can be expressed as a short pulse with frequency components in each sub-band. Conversely, orthogonality between the different wavelet sub-spaces ensures a quasi decorrelation of the speckle noise, which behaves as a random distributed noise identically distributed within each scale.

According to the wavelet theory [18, 12], the DWT of a signal $x[n]$ is computed by passing it through a series of filters. First the signal is passed through both a low-pass and a high-pass filter with impulse responses $w_l[n]$ and $w_h[n]$, resulting in two convolutions:

$$y_L[n] = \sum_{p=-\infty}^{\infty} x[n]w_l[n-p] \tag{7.1}$$

$$y_H[n] = \sum_{p=-\infty}^{\infty} x[n]w_h[n-p] \tag{7.2}$$

The outputs are the *detail coefficients* (from the high-pass filter) and the *approximation coefficients* (from the low-pass filter). However, according to the Nyquist theory, since half of the frequencies of the signal have now been removed, half of the samples can be discarded. Then the filter outputs are actually defined as follows:

$$y_L[n] = \sum_{p=-\infty}^{\infty} x[n]w_l[2n-p] \tag{7.3}$$

$$y_H[n] = \sum_{p=-\infty}^{\infty} x[n]w_h[2n-p] \tag{7.4}$$

The decomposition is repeated to further increase the frequency resolution. Then, the approximation coefficients are decomposed and down-sampled. This can be modeled as a binary tree where nodes are sub-spaces with different time-frequency localization. This tree is known as a *filter bank*.

The block diagram in Figure 7.2 describes this decomposition for a 2D SAR image. In this case, the same processing described before is applied first at each row and then at each column. Then, $[2 \downarrow 1]$ and $[1 \downarrow 2]$ are the column down-sampling operator and the row down-sampling operator, respectively.

Three wavelets can then be defined (HL,LH,HH) each of them extracting image details for a given orientation, namely, D_{j_v}, D_{j_h} and D_{j_d}. The low-pass filtered version of the original image is denoted as LL.

At each iteration and according to their orientation, the 2D-DWT performs a separation of the different frequency components of the original image. More specifically, coefficients of large values in different sub-bands correspond to

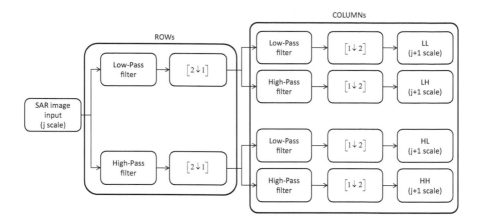

Figure 7.2 2D-DWT block-scheme for the image decomposition

vertical, horizontal and edges. According to [19], the DWT with Haar coefficients was chosen among the different families of wavelet transform because it is appropriate for detection of point-like targets.

The idea to use an intra-scale correlation computed on the wavelet decomposition of the image has been proposed by Kuo and Chen [10] with the intent of enhancing the edge-like feature and wakes prior to apply the Radon or the Hough transform. A similar approach has been used here to improve even more the ships detection.

These considerations suggest to use a *wavelet correlator* (WCOR), which consists of calculating the modulus of the detailed images resulting from each scale and applying the detection directly in the wavelet domain with a thresholding operation. As a drawback, the sizes of the sub-images at the scale j are reduced by a factor of 2^{-j} compared to their original sizes.

To extract features from an image by combining different scales, the sizes must be the same across different scales. Since the down-sampling will result in a fine to coarse resolution scale leading to images with different sizes, the detailed images cannot be used directly. To account for this, first the modulus of the detailed image at the jth is defined as follows:

$$M_j(x_1, x_2) = \sqrt{(|D_{j_h}(x_1, x_2)|^2 + |D_{j_v}(x_1, x_2)|^2 + |D_{j_d}(x_1, x_2)|^2)} \qquad (7.5)$$

where the three detailed images, $D_{j_h}(x_1, x_2)$, $D_{j_v}(x_1, x_2)$ and $D_{j_d}(x_1, x_2)$ in the horizontal, vertical and diagonal directions, respectively, are generated for each resolution scale and (x_1, x_2) represents the spatial coordinates of a pixel. Then, the WCOR function $R(l, x_1, x_2)$ may be defined for several adjacent scales as follows:

$$R(l_s, x_1, x_2) = \prod_j \left[\uparrow 2^{j-1}\right] M_j(x_1, x_2) \qquad (7.6)$$

where l_s represents the number of scales involved in the direct correlation and $\left[\uparrow 2^{j-1}\right]$ means the resampling of $M_j(x_1, x_2)$ with $j > 1$ to the same size as M_1.

The choice of l_s depends on the degree of noise associated with the features to be detected and also on the spatial resolutions of the original SAR image. A likely value for l_s is 3. Pixels of a ship are closely related in terms of intensity and statistical characteristic, since they belong to the same structure. As a consequence, the product in Equation (7.6) emphasizes these relations among different scales. On the contrary, sea clutter and speckle pixels remain randomly distributed. Since the 2D-WDT tend to be sparse, the noise power is reduced due to the projections. Moreover, the correlation provides reduced intensity values due to the low probability of coincidence of related coefficients.

7.1.2   DETECTOR

The detector is mainly composed of two steps: (i) the S-detector, which aims at identifying the areas where targets are likely present, and (ii) a CFAR-based detector applied only on those regions identified by the pre-detector. A block scheme of the proposed algorithm is shown in Figure 7.3

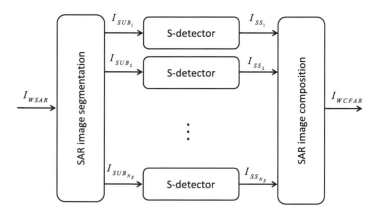

Figure 7.3   Detector block-scheme

7.1.2.1   S-Detector

The wavelet correlator produces an enhanced image so both targets and land areas are better identifiable from a visual point of view. However, the size of such an image is the same as the original size. As a consequence, the direct application of a standard CFAR-based algorithm may result in a high computational burden.

Then, the idea is to identify regions of interest in which targets of interest are likely present instead of applying the detection algorithm to the whole SAR image.

Such a step can be interpreted as pre-detection step and in this work is based on the *significance*, which is called *S-detector*.

The significance parameter is a measure of the image contrast and then a measure of the goodness of the image quality. It is defined as follows:

$$S = \frac{max|I_{WCOR}| - m_I}{\sigma_I} \tag{7.7}$$

where I_{WCOR}, m_I and σ_I represent the SAR image after the WCOR processing, its mean value and its standard deviation.

First, the whole SAR image is split into sub-images (blocks). The size of each sub-image is chosen so as to have a good trade-off between the processing time, the detection performances and the size of the targets of interest. Then, the significance is computed on all the sub-images. The significance will exhibit a high value in the presence of targets and a low value otherwise.

The sub-images where the significance exceeds the detection threshold T are considered as candidates for containing targets and they will be processed by the W-CFAR at the subsequent step. The detection is made by comparing the value of the significance of the block under test against a detection threshold T. The threshold is computed by assuming a probability density function fitting the histogram of the S values in clutter only conditions and by fixing a desired probability of false alarm P_{FA}. More precisely, the probability of false alarm is defined as follows:

$$P_{FA} = Pr\{s > T|H_0\} = \int_T^\infty g(s|H_0)ds = $$
$$1 - \int_{-\infty}^T g(s|H_0)ds = 1 - G(T|H_0) \tag{7.8}$$

where $g(s|H_0)$ is the probability density function, *pdf*, of the significance, s, conditioned to the hypothesis H_0, namely clutter only condition. Then the threshold can be found as follows:

$$T = G^{-1}(1 - P_{FA}|H_0) \tag{7.9}$$

where $G(s|H_0)$ is the cumulative density function of s conditioned to the hypothesis H_0. Since the the random variable s represents the maximum value of the image intensity, the evaluation value theory (EVT) can be exploited in this case to model the probability density function of s.

The EVT is usually compared with the central limit theorem. However, while the central limit theorem deals with the sum of random variables,

EVT deals with the maximum of random variables. The central limit theorem states that the statistical distribution of the sum of independent and identically distributed random variables converges to a given well-known distribution, namely the normal one. Similarly, Fisher, Tippett and Gnedenko [6, 8] demonstrate that the maximum of a sequence of independent and identically distributed random variables converges to a given distribution family. A single representation is provided by the generalized extreme value (GEV) distribution [9].

The GEV distribution has been formulated to model a family of three standard *pdf* functions. This can be accomplished by introducing a parameter b so that:

$b = 0$ corresponds to the Gumbel distribution Λ
$b = a^{-1} > 0$ corresponds to the Fréchet distribution Φ_a
$b = a^{-1} < 0$ corresponds to the Weibull distribution Ψ_a

where

$$\Lambda(x) = exp\{-exp\{-x\}\} \text{ where } x \in \mathbb{R} \tag{7.10}$$

where $exp\{x\} = e^x$.

$$\Phi_a(x) = \begin{cases} 0 & x < 0 \\ exp\{-x^{-a}\} & x \geq 0 \end{cases} \tag{7.11}$$

$$\Psi_a(x) = \begin{cases} exp\{-(-x)^a\} & x < 0 \\ 0 & x \geq 0 \end{cases} \tag{7.12}$$

where x in this context represents a random variable realization. The standard form of the GEV distribution can be written as follows:

$$g(x) = \begin{cases} exp\{-(1+bx)^{1/b}\} & \text{if } b \neq 0 \\ exp\{-exp\{-x\}\} & \text{if } b = 0 \end{cases} \tag{7.13}$$

where $1 + bx > 0$.

Hence, the support of the function $g(x)$ corresponds to:

$$\begin{cases} x > -b^{-1} & \text{if } b > 0 \\ x < -b^{-1} & \text{if } b < 0 \\ x \in \mathbb{R} & \text{if } b = 0 \end{cases} \tag{7.14}$$

The location-scale family, $g(x)$, can then be defined by substituting the argument x with $(x - \mu_g)/\sigma_g$ where $\mu_g \in \mathbb{R}$ and $\sigma_g > 0$.

Then, Equation (7.14) becomes:

$$\begin{cases} x > \mu_g - \frac{\sigma_g}{b} & \text{if } b > 0 \\ x < \mu_g - \frac{\sigma_g}{b} & \text{if } b < 0 \\ x \in \mathbb{R} & \text{if } b = 0 \end{cases} \tag{7.15}$$

The two parameters, μ_g and σ_g are the location and the unknown scale parameters. Such parameters are estimated via the maximum likelihood theorem.

It can be proven that, by assuming the GEV distribution, the threshold T can be found as follows:

$$
T = \begin{cases} a_1 - \frac{a_2}{b}\left[1 - ln^{-b}\frac{1}{1-P_{FA}}\right] & \text{for } b \neq 0 \\ a_1 - b\left[1 - ln^{-b}\frac{1}{1-P_{FA}}\right] & \text{for } b = 0 \end{cases} \tag{7.16}
$$

where a_1 and a_2 represent the location and scale parameters of the probability density function and therefore are estimated from clutter only regions, while b is the GEV distribution parameter.

7.1.2.2 W-CFAR

After the *S-detector*, all the sub-images whose value exceeds the threshold T are processed by the W-CFAR detector [17, 16]. The W-CFAR processor searches for pixels that are unusually bright compared with those of the surrounding area. The comparison is carried out by means of a threshold which is computed via the formula in Equation (7.17), for a given probability of false alarm P_{FA}:

$$
P_{FA} = \int_{T_r}^{\infty} g_{A_{ref}}(x)\, dx \tag{7.17}
$$

where $g_{A_{ref}}(x)$ is the probability density function (*pdf*) estimated from the pixels in the reference area A_{ref}, which should only contain clutter pixels and T_r is the threshold which gives the desired P_{FA}. As can be easily inferred from Equation (7.17), the threshold depends on the statistics of the surrounding area and is adaptively computed to ensure a constant false alarm rate. The clutter statistics are usually analyzed by exploiting an area around the CUT, as depicted in Figure 7.4.

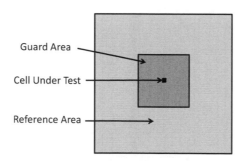

Figure 7.4 Guard area and reference area definition

It consists of three different areas: (i) the investigated CUT or pixel, (ii) the guard area, and (iii) the reference area. The reference area contains the reference cells or pixels that will be used to estimate the clutter statistics. The guard area is instead chosen to be sure that pixels belonging to the target will not be used to estimate the clutter statistics. In fact, when a portion of the target falls into the reference area, the clutter only assumption decays, thus affecting the estimates of the clutter statistics. The guard area is chosen as an area adjacent to the CUT and its sizes are in some way linked to the maximum expected target size. The size of the reference cells is instead linked to both the desired accuracy of the clutter statistics estimates and the non-stationarity of the clutter. In fact, under the hypothesis of identically distributed random variables, as a rule of thumb, the more the reference cells the better the estimates. However, the clutter is usually non-stationary; then, the reference area should be small enough to be sure that clutter can be considered stationary within the reference area.

The most widely accepted model to statistically characterize the sea clutter in SAR images is the $K - distribution$. The estimation of the parameters for this type of distribution can be very time consuming due to the time needed to invert both the Bessel and the Gamma functions. This is also true when using the efficient and consistent but still approximate numerical method. To overcome this problem, the *LogNormal* distribution can be used, which can be more easily managed. The *LogNormal pdf* is in Equation (7.18)

$$g_{LN}(x) = \frac{1}{x\sqrt{2\pi\sigma_x^2}} \exp\left[-\frac{(ln(x) - \mu_x)^2}{2\sigma_x^2}\right] \qquad (7.18)$$

where μ_x and σ_x are, respectively, the mean and the standard deviation (namely the location and scale parameters) of the associated normal distribution and $ln(x) = log_e(x)$. Once a model for the first order *pdf* of the sea clutter has been fixed, its unknown parameters must be estimated by using clutter only pixels in the nearby of the CUT. In the hypothesis of i.i.d. clutter samples, the clutter statistics can be estimated via the maximum likelihood method as shown in Equation (7.19) and (7.20).

$$\hat{\mu}_x = \frac{1}{N_{ref}} \sum_{k=0}^{N_{ref}-1} ln(x_k) \qquad (7.19)$$

$$\hat{\sigma}_x^2 = \frac{1}{N_{ref}} \sum_{k=0}^{N_{ref}-1} (ln(x_k) - \mu_x)^2 \qquad (7.20)$$

where N_{ref} is the number of pixels x_k in the reference area A_{ref}.

By using the estimates in Equation (7.19) and (7.20) and by considering the *LogNormal* distribution in Equation (7.18) the threshold can be found as in Equation (7.21):

$$T_r = \exp\left[\Phi^{-1}\left(1 - P_{FA}\right) \cdot \hat{\sigma}_x + \hat{\mu}_x\right] \tag{7.21}$$

where $\Phi^{-1}(\cdot)$ is the inverse of the normal cumulative distribution function (CDF) defined in Equation (7.22)

$$\Phi(x) = \frac{1}{\sqrt{2\pi}} \int_{\infty}^{x} e^{-\frac{t^2}{2}} \, dt \tag{7.22}$$

7.1.3 POST-PROCESSING

Large ships in high spatial resolution (of the order of 1 m) SAR images may occupy several resolution cells. As a consequence, a certain number of segments or groups of pixels which actually belong to the same target may be detected and a single target may be divided into different groups of pixels. The clustering algorithm is then applied to avoid such a situation and merge those groups of pixels.

A well-established definition of the clustering can be given as follows: *Clustering is the task of grouping a set of objects in such a way that objects in the same group have similar characteristics.*

For our purposes, for example, the segments or group of pixels are grouped together by taking into account the relative distance (in terms of image pixels) between the centroids of each segment. Other rules could be considered depending on the available SAR system. As an example, when a fully polarimetry or a multi-channel SAR system is available, the polarimetric signature or the interferogram could be used in conjunction with the distance to improve the false alarm rejections and the clustering operation.

The clustering processing can be thought of as composed of three main steps: (i) a *clean* morphological filter, (ii) the labeling and fusion step, and (iii) the removal of artifacts or sidelobes of strong scatterers.

Figure 7.5 Post-processing block-scheme

The *clean* algorithm is used to reduce the number of false alarms. Then, isolated group of pixels whose size cannot be compatible with those of a ship will be discarded.

After the *cleaning* operation, a labeling process is applied to label connected groups of pixels. Moreover, the centroid of each group of pixels, also called segment, is computed.

After this stage, segments likely belonging to the same target are grouped together. A simple rule is based on the relative distance among centroids of different segments. Then, segments whose centroids are far apart less than a certain range and cross-range distance are declared belonging to the same target. The segments are first ordered based on their range and cross-range sizes. Then, starting from the largest one, the distances between its centroids and those of the others segments are computed. The nearest segment is then fused with the large segment and a new centroid is computed. This procedure is then repeated until there are segments whose distance from the reference one is less than the pre-defined threshold. The remaining segments are then analyzed again in the same way to search for other targets. The *closing* morphological filter is finally applied to each detected target so as to fill the holes among segments constituting a target.

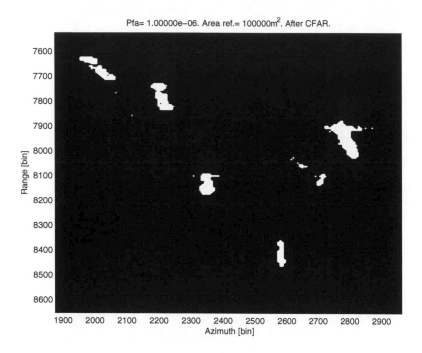

Figure 7.6 Example of a black and white radar image at the output of the detector. Pixels in white have passed the threshold

To further reduce the probability of false alarm, another *cleaning* operation can be performed. Finally, the remaining small isolated segments are deleted. Moreover, ground bright scatterers must be identified as they may cause high sidelobes in the sea region and consequently affect the detection performances. Then a quite simple rule based on comparing range/cross-range distance of segments with the same cross-range/range centroids coordinates is applied. Figure 7.6 and Figure 7.7 show a pictorial representation of such a step. Specifically, Figure 7.6 shows the result of the W-CFAR detector, while

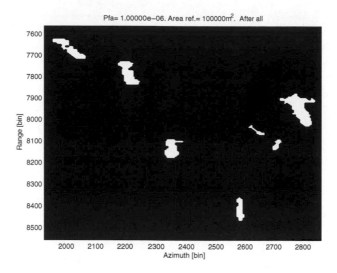

Figure 7.7   Image in Figure 7.6 at the output of the post-processing

Figure 7.7 depicts the result of the post-processing step. The image refers to a portion of a larger SAR image covering the area nearby Istanbul harbour (Figure 7.8).

7.2   EXPERIMENTAL RESULTS

Some results are shown in this section to demonstrate the feasibility of the described algorithm. More specifically, the algorithm has been tested on space-borne SAR images acquired with COSMO-SkyMed system and on an airborne EMISAR data.

COSMO-SkyMed (COnstellation of small Satellites for Mediterranean basin Observation) is the largest Italian investment in space systems for earth observation commissioned and founded by the Italian Space Agency (ASI) and the Ministry of Defence (MoD). COSMO SkyMed is a dual-use (for civil and military applications) end-to-end system for earth observation and has been conceived for a wide range of applications such as risk management, scientific and commercial applications, homeland security, defense and intelligence applications. COSMO-SkyMed constellation consists of four low earth orbit mid-size satellites, each equipped with a high-resolution SAR system operating at X-band. As different applications ask for different requirements in terms of both SAR image spatial resolutions and SAR swath, COSMO-SkyMed has been equipped by a multi-model SAR system, in which each mode differs from the others in term of SAR image spatial resolutions and SAR swath. More specifically three different modalities has been defined:

1. Spotlight mode, which provide SAR images of small swath (of the order of 10 km) with high spatial resolutions (of the order of 1 m)
2. Two stripmap modes (*Himage* and *PingPong* modes), conceived for SAR images of medium size (tenth of km) and spatial resolutions (of the order of 10 m). One such mode, the *PingPong* one, provides SAR images in two different polarizations by alternating the signal polarization between two of the possible ones, i.e., HH, VV, HV, VH
3. Two ScanSAR modes (*WideRegion* and *HugeRegion*) for medium to coarse spatial resolutions (of the order of 100 m) and large swath (of the order of 100 km)

A ground segment is also present which is responsible for the operations and control of the entire system including the generation and dissemination of the final products. Different products can be delivered by the ground segment, among which we are interested in the *Level 1A* data which is a single-look complex slant product, in other words, the raw data focused in slant-range/azimuth plane, which is the sensor natural acquisition projection.

More specifically, two spotlight SAR images have been considered hereinafter to test the detection algorithm.

Another SAR image acquired by means of the EMISAR SAR system has been also analyzed. Details about the EMISAR system can be found in [2].

7.2.1   CASE STUDY #1

The first case study comprehends a SAR image acquired on April 15, 2008 which covers the area of Istanbul. Figure 7.8 shows the SAR image.

7.2.1.1   Pre-Processing Results

In this section, the results of the pre-processing algorithm are shown. Figure 7.9 shows the SAR image after the wavelet correlator. As can be seen by comparing this image with that in Figure 7.8, the wavelet correlator has well performed the speckle reduction allowing for a better understanding of both the sea-land and the sea-target edges.

Before going on and with the aim to reduce the computation time, the land areas should be identified and separated from the sea areas, so the detection algorithm will be applied only on sea regions.

Then a simple algorithm based on the analysis of the histogram of the SAR image intensity has been used. More specifically, the histogram computed on the SAR image after wavelet correlator is shown in Figure 7.10. As expected, the histogram is sketchily composed of two peaks, one corresponding to lower intensity values related to the sea pixels, the other corresponding to higher intensity values related to both land and targets pixels. Then a threshold can be set to separate these two regions. However, as can be noted by carefully inspecting Figure 7.10, even if the two peaks can be recognized, the threshold separating them cannot be easily identified.

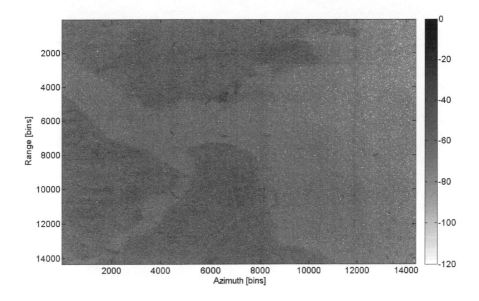

Figure 7.8 SAR image of the area nearby Istanbul harbour. Courtesy of ASI

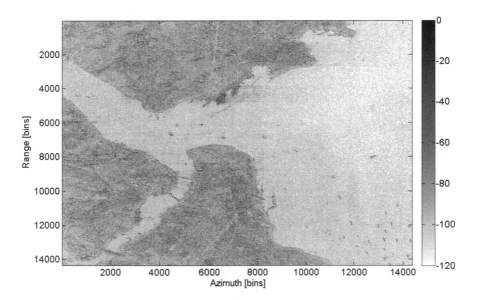

Figure 7.9 SAR image after wavelet correlator

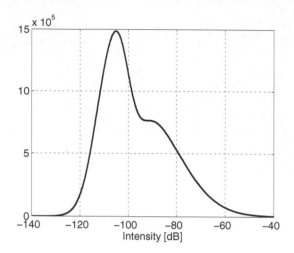

Figure 7.10   Histogram computed from image in Figure 7.9

Then, to improve the sea-land separability, a low-pass spatial filter has been applied over the same image thus obtaining the image in Figure 7.11. Such filter allows for a better reduction of clutter and noise peaks thus improving the sea-land separation, as can be easily noted by means of a visual inspection.

The histogram computed on the image in Figure 7.11 is shown in Figure 7.12. As can be seen, although the intensity levels are different, the two peaks are more easily identified. A threshold can now be computed as the value corresponding to the local minimum among two local maxima.

When such threshold is applied on the SAR image in Figure 7.11, the sea-land mask is obtained and is shown in Figure 7.13. Then, the contour separating the land area from the sea can be easily extracted and applied to the SAR images in Figure 7.8 and Figure 7.9. The sea-land separation step allows to speed up the detection algorithm as it is now applied only to sea regions.

7.2.1.2   Detection Results

As detailed in Section 7.1, to further speed up the detection of targets, the detection algorithm has been thought of as composed of two steps, a pre-detection step based on the significance parameter, and an adaptive detection algorithm based on the use of an adaptive threshold.

Before showing the results of the pre-detection step, the statistical analysis of the significance parameter should be performed first to confirm the hypothesis made in the previous section.

For this purpose, a sea only clutter region has been identified as shown in Figure 7.14. A sliding window of about 40 × 40 pixels has also been used

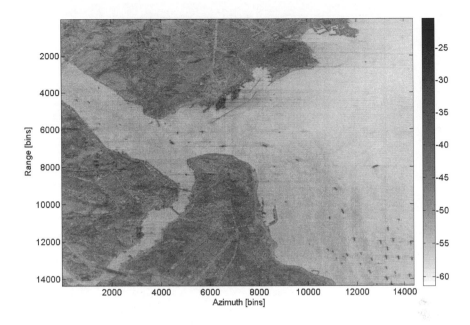

Figure 7.11 SAR image after a low-pass smoothing filter

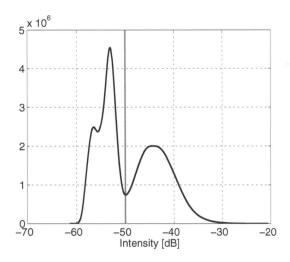

Figure 7.12 Histogram computed from image in Figure 7.11

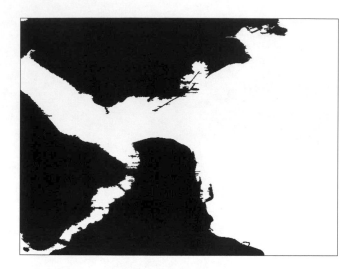

Figure 7.13 Sea-land mask

to analyze the statistical properties of the random variable S, namely the significance.

To estimate the significance *pdf*, its histogram is calculated first. By defining $\mathcal{K}(s)$ as the histogram, the probability that a significance value falls into an interval Δs is

$$
\begin{aligned}
Pr_s =&Pr\left\{s - \frac{\Delta s}{2} < S < s + \frac{\Delta s}{2}\right\} \\
&\int_{s-\frac{\Delta s}{2}}^{s+\frac{\Delta s}{2}} f(s)\ ds \simeq f(s)\Delta s \simeq \frac{\mathcal{K}(s)}{N_S}
\end{aligned}
\tag{7.23}
$$

where N_S is the number of significance samples. Thus, an estimate of the significance *pdf* can be derived from Equation (7.23), as follows:

$$
f(s) \simeq \frac{1}{N_S} \cdot \frac{\mathcal{K}(s)}{\Delta s}
\tag{7.24}
$$

The significance *pdf* is depicted in Figure 7.15, where some well-known distributions fitting that of the significance are also plotted. As can be noted, the distribution that best approximates the *pdf* significance is, as expected, the GEV one. The estimated statistical unknown parameters of the GEV distribution are listed in Table 7.1.

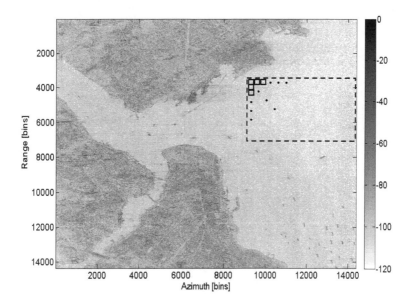

Figure 7.14 Pictorial representation of the procedure to evaluate the random variable S

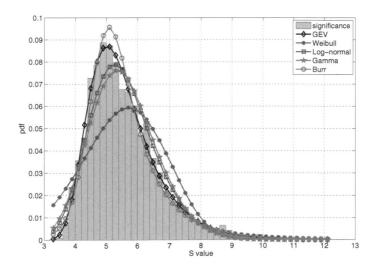

Figure 7.15 Probability density function of the significance parameter computed from Figure 7.9

Table 7.1

GEV statistical parameters fitting the significance pdf

b	a_2	a_1
0.05869	0.911819	5.82404

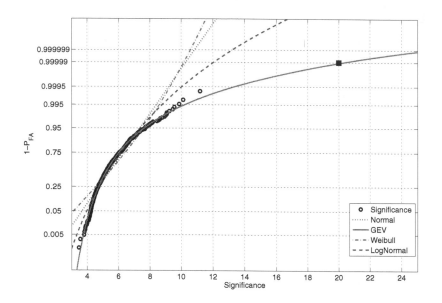

Figure 7.16 Cumulative density function of the significance parameter computed from Figure 7.9 and adaptive threshold

Once a distribution fitting the significance estimated *pdf* has been found, the threshold can be derived from Equation (7.16). By assuming the GEV distribution and by substituting the parameters in Table 7.1 in Equation (7.16), the threshold $T \simeq 20.822$ can be found, which corresponds to a probability of false alarm equal to $P_{FA} = 10^{-5}$.

The same threshold can be found by representing the significance cumulative density function (CDF), as shown in Figure 7.16. Then, the blocks where the significance value exceeds the threshold likely contains one or more targets and are then processed by the adaptive detection algorithm.

The pre-detection algorithm is then applied to the whole SAR image. The SAR image has been divided into blocks of 800 × 800 pixels with an overlap of 40%. The significance values are shown in Figure 7.17, where the threshold is also shown. Blocks for which significance is equal to 0 correspond to land areas.

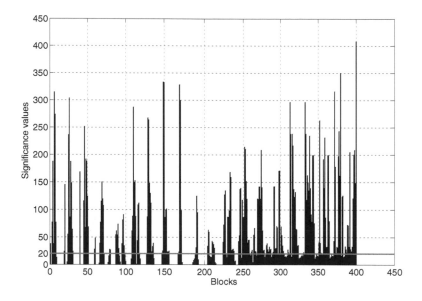

Figure 7.17 Significance values

Figure 7.18 shows the SAR image where both the land pixels and the blocks with significance below the threshold have been put equal to zero. Once the pre-detection is performed, the blocks where significance exceeds the threshold are processed by the adaptive detection algorithm. To analyze the detector performance a ground truth map has been created via visual inspection.

All the results shown hereinafter have been obtained by using a nominal probability of false alarm $P_{FA} = 10^{-6}$ to set the detector threshold. However, before showing the detection results, the statistical analysis of the clutter must be performed first. Then, as done previously, an area free of targets and land has been selected from the SAR image in Figure 7.9. Then, the histogram of sea clutter intensity has been computed as shown in Figure 7.19 where some well-known distributions fitting that of the sea clutter are also plotted. As can be noted, the distribution that best approximates the clutter distribution is, as expected, the *Log-Normal* one. Since the clutter is usually non-stationary, its statistics may change. Then, for each CUT, the algorithm chooses a reference area of pre-defined size and computes the statistical parameters of the Log-Normal distribution. Once the model has been fixed, the threshold is then derived for a fixed P_{FA}.

Figure 7.20 shows the results of the adaptive detector where targets are represented in black while the coast line has been represented with a solid gray line.

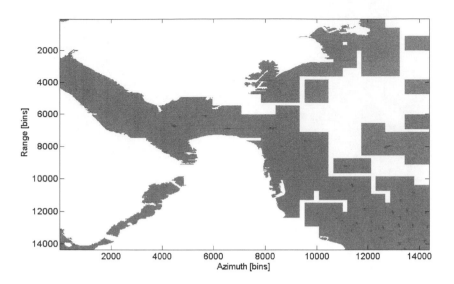

Figure 7.18   Results of the significance analysis

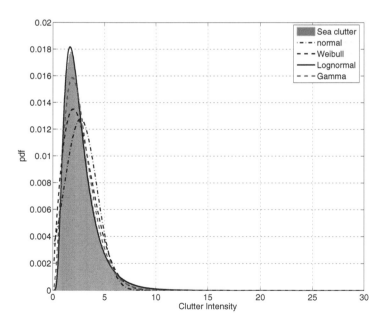

Figure 7.19   Probability density function of the sea clutter estimated from SAR image in Figure 7.9

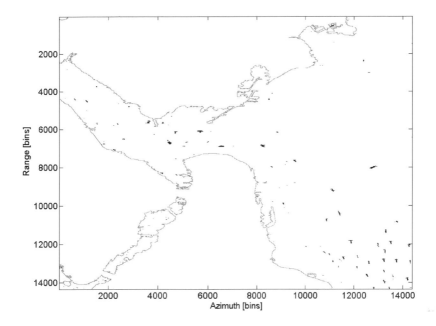

Figure 7.20   Detection results

After the post-processing algorithm, some false alarm and some artifacts have been removed from the detection map, thus providing the results in Figure 7.21. To get such results, a reference area of $A_{ref} = 0.15\ km^2$ has been considered around the CUT. Such value has been found empirically and by maximizing the probability of detection, P_D and minimizing the probability of false alarm, P_{FA}. Both P_D and P_{FA} have been computed from the detection map in Figure 7.21 as follows:

$$\hat{P}_D = \frac{N_D}{N_{SAR}} \simeq 0.941 \tag{7.25}$$

$$\hat{P}_{FA} = \frac{N_{FA}}{N_{SAR}} \simeq 0.6\,10^{-4} \tag{7.26}$$

where N_D, N_{FA} and N_{SAR} are, respectively, the number of detected pixels, false alarm and the total number of investigated pixels.

7.2.2   CASE STUDY #2

The second case study comprehends a SAR image acquired on April 23, 2008 which covers the area of Messina harbor. Figure 7.22 shows the SAR image.

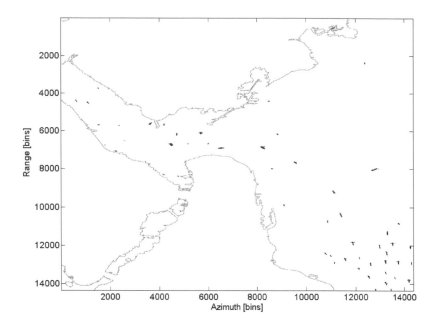

Figure 7.21 Detection results after post-processing

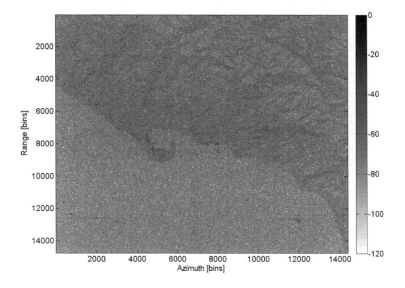

Figure 7.22 SAR image covering the area of Messina, Italy. Courtesy of ASI

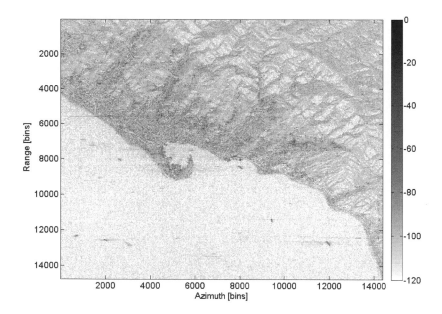

Figure 7.23   SAR image after wavelet correlator

7.2.2.1   Pre-Processing Results

Figure 7.23 shows the SAR image after the wavelet correlator. As done previously, the sea-land separation has been performed before detection based on the histogram of the SAR image intensity. The histogram computed on the SAR image in Figure 7.23 is represented in Figure 7.24. It can be noted that even if two peaks can be identified which correspond to the sea and land pixels, it is difficult to identify a threshold that can easily separate them. Then, a spatially low-pass filter has been applied to the SAR image in Figure 7.23 thus obtaining the SAR image in Figure 7.25 where the coast line is even more evident. The histogram computed on such SAR image is shown in Figure 7.26. A threshold can now be easily identified which separates sea pixels from land pixels as shown in Figure 7.27.

7.2.2.2   Detection Results

After the pre-processing stage, the pre-detector is applied to search for the region where targets are likely present so as to further speed up the detection. As done previously, the statistical analysis of the significance has been carried out first. Then, a sea clutter only region has been identified and a sliding window of 40×40 pixels has been used to analyze the statistical properties of the random variable S. To accomplish this task, the histogram of the sea clutter

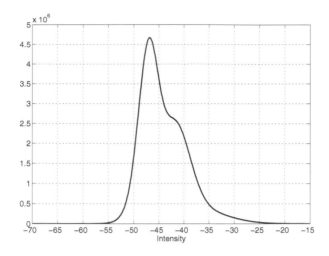

Figure 7.24 Histogram computed from image in Figure 7.23

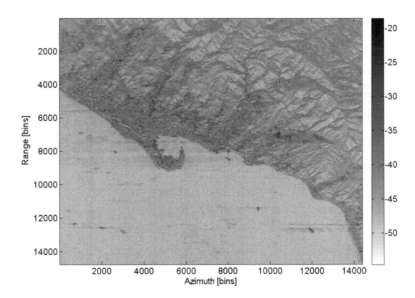

Figure 7.25 SAR image after wavelet correlator and after low-pass smoothing filter

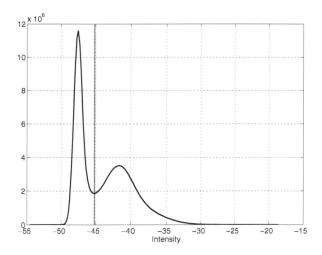

Figure 7.26 Histogram computed from image in Figure 7.25

Table 7.2
GEV statistical parameters fitting the significance pdf

b	a_2	a_1
0.040449	0.953012	6.23434

region has been computed and is represented in Figure 7.28 where some well-known distribution fitting the significance *pdf* are also shown. As expected, the distribution that best approximates the *pdf* significance is the GEV one. The estimated statistical unknown parameters of the GEV distribution are listed in Table 7.2.

Once a distribution fitting the significance one has been found, the threshold can be derived from Equation (7.16). More specifically, to get a probability of false alarm equal to $P_{FA} = 10^{-5}$ a threshold equal to $T \simeq 20.2085$ should be used. The same result can be found via the significance *CDF*, as shown in Figure 7.29. Once the threshold has been fixed, the pre-detection algorithm is applied to the whole SAR image. In this case the SAR image has been divided into blocks of 800 × 800 pixels with an overlap of 40%. The block size is chosen based on typical target size. The significance values are shown in Figure 7.30. The significance relative to land areas has been forced equal to zero. Figure 7.31 shows the SAR image where both the land pixels and the blocks for which the significance is lower than the threshold have been put equal to zero.

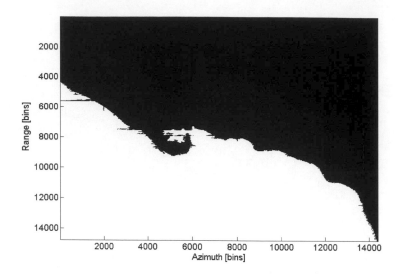

Figure 7.27 Sea-land separation results

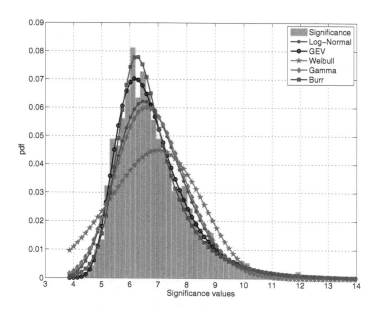

Figure 7.28 Probability density function of the significance parameter computed from Figure 7.23

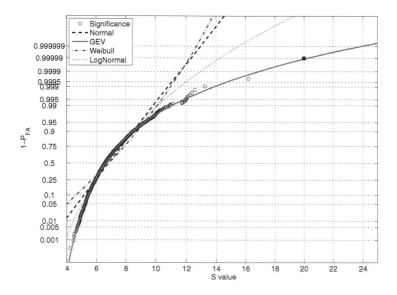

Figure 7.29 Cumulative density function of the significance parameter computed from Figure 7.23 and adaptive threshold

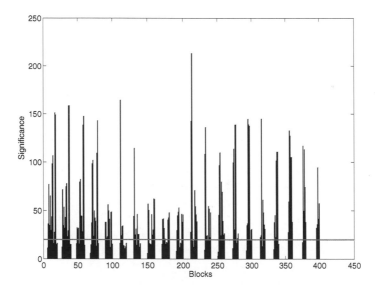

Figure 7.30 Significance values

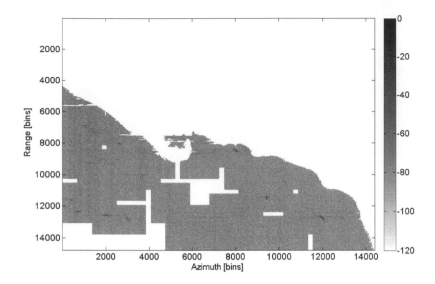

Figure 7.31   Results of the significance analysis

After the pre-detection, the blocks where significance exceeds the threshold are then processed by the adaptive detection algorithm. To test the detection algorithm performances, a ground truth map has been created via visual inspection of the SAR image. The results shown hereinafter have been obtained by fixing a nominal probability of false alarm $P_{FA} = 10^{-6}$. However, also in this case, the statistical analysis of the clutter has been performed first. An area free of targets and land pixels has been selected from the SAR image in Figure 7.23. Then, the histogram of the sea clutter intensity has been computed as shown in Figure 7.32. As can be noted, the distribution that best fits the clutter *pdf* is, again, the *LogNormal* one.

Figure 7.33 shows the results of the adaptive detector where targets are represented in black while the coast line is represented with a solid gray line. After the post-processing step, some false alarms and some artifacts have been removed, thus leading to the results in Figure 7.34. To get such results a reference area of $A_{ref} = 0.15 \ km^2$ has been considered.

As done previously, such value has been found empirically and by maximizing the probability of detection, P_D and minimizing the probability of false alarm, P_{FA}. Both P_D and P_{FA} have been computed from the detection map in Figure 7.34 as follows:

$$\hat{P}_D = \frac{N_D}{N_{SAR}} \simeq 0.981 \tag{7.27}$$

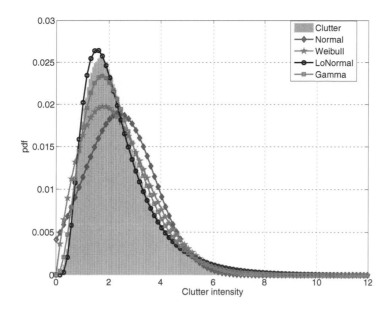

Figure 7.32 Probability density function of the sea clutter intensity computed from Figure 7.23

$$\hat{P}_{FA} = \frac{N_{FA}}{N_{SAR}} \simeq 2.53\,10^{-4} \qquad (7.28)$$

where N_D, N_{FA} and N_{SAR} are, respectively, the number of detected pixels, false alarms and the total number of investigated pixels.

7.2.3   CASE STUDY #3

The third case study uses an EMISAR airborne polarimetric data set covering the area of the Storebaelt bridge. Figure 7.35 shows the SAR image. Some important parameters of such SAR data are listed in Table 7.3.

Table 7.3

Parameters of interest of the Storebaelt bridge SAR image

Central frequency, f_0	$5.3\,GHz$
Azimuth spacing	$0.749\,m$
Range spacing	$1.499\,m$
Bandwidth	$100\,MHz$

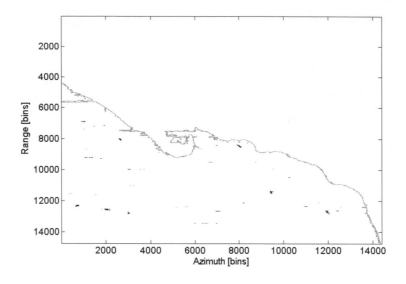

Figure 7.33 Detection results before post-processing

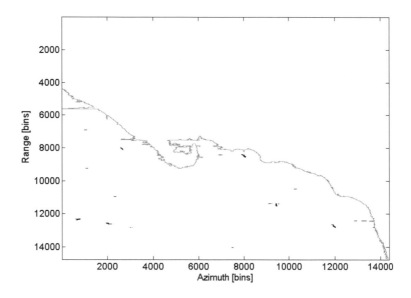

Figure 7.34 Detection results after post-processing

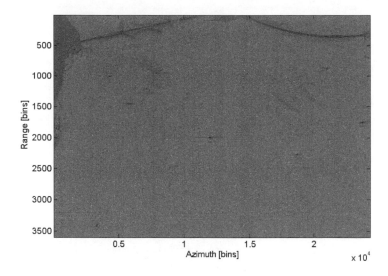

Figure 7.35 SAR image

7.2.3.1 Pre-Processing Results

Figure 7.36 shows the SAR image after the wavelet correlator. As done previously, the sea-land separation has been performed before detection based on the histogram of the SAR image intensity. A spatially low-pass filter has been applied to the SAR image in Figure 7.36 thus obtaining the SAR image in Figure 7.37 where the coast line is even more evident. The histogram computed on such SAR image is shown in Figure 7.38. A threshold can now be easily identified which separates sea pixels from land pixels as shown in Figure 7.39.

7.2.3.2 Detection Results

After the pre-processing stage, the pre-detector is applied to search for the region where targets are likely present so as to further speed up the detection. As done previously, the statistical analysis of the significance has been carried out first. Then, a sea clutter only region has been identified and a sliding window of 50 × 50 pixels has been used to analyze the statistical properties of the random variable S. To accomplish this task its histogram has been computed and is represented in Figure 7.40 where some well-known distribution fitting the significance *pdf* are also shown. As expected, the distribution that best approximates the *pdf* significance is the GEV one. The estimated statistical unknown parameters of the GEV distribution are listed in Table 7.4.

Once a distribution fitting the significance one has been found, the threshold can be derived from Equation (7.16). More specifically, to get a probability

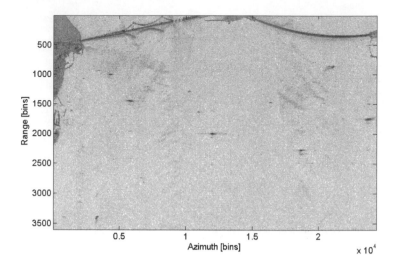

Figure 7.36 SAR image after wavelet correlator

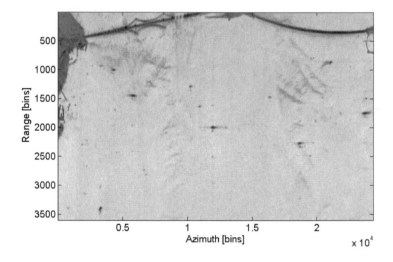

Figure 7.37 SAR image after wavelet correlator and after low-pass smoothing filter

---

Table 7.4

GEV statistical parameters fitting the significance pdf

b	a_2	a_1
0.03586	0.9097	5.5369

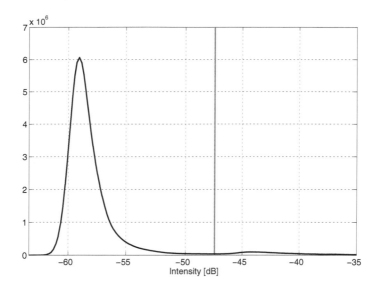

Figure 7.38   Histogram computed from image in Figure 7.37

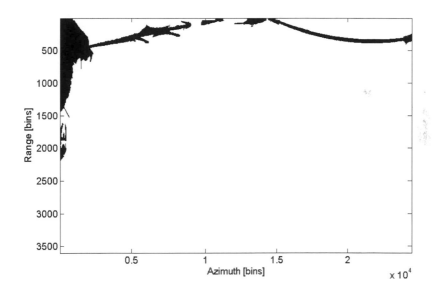

Figure 7.39   Sea-land separation results

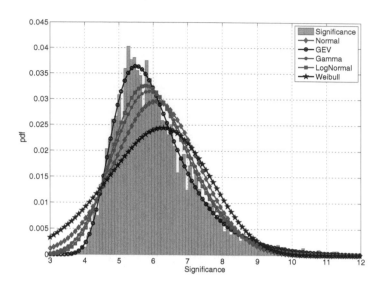

Figure 7.40 Probability density function of the significance parameter computed from Figure 7.36

of false alarm equal to $P_{FA} = 10^{-5}$ a threshold equal to $T \simeq 18.5$ should be used. The same result can be found via the significance *CDF*, as shown in Figure 7.41. Once the threshold has been fixed, the pre-detection algorithm is applied to the whole SAR image. In this case the SAR image has been divided into blocks of 600×600 pixels with an overlap of 40%. The block size is chosen based on typical size of targets we are interested in. The significance values are shown in Figure 7.42. The significance relative to land areas has been forced equal to zero.

After the pre-detection, the blocks where the significance exceeds the threshold are then processed by the adaptive detection algorithm. To test the detection algorithm performance a ground truth map has been created via visual inspection of the SAR image. The results shown hereinafter have been obtained by fixing a nominal probability of false alarm $P_{FA} = 10^{-6}$. However, also in this case, before showing the results, the statistical analysis of the clutter has been performed first. An area free of targets and land pixels has been selected from the SAR image in Figure 7.36. Then, the histogram of the sea clutter intensity has been computed as shown in Figure 7.43. As can be noted, both the *LogNormal* distribution and the *GEV* distribution well fit the clutter *pdf* distribution. However, as a closed form solution for the estimation of the GEV statistics does not exist, the computation of such parameters may be time consuming. Conversely, the statistics of the *LogNormal* distribution may be found via the formula in Equation (7.19) and (7.20).

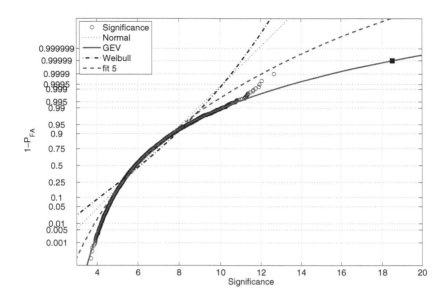

Figure 7.41 Cumulative density function of the significance parameter computed from Figure 7.36 and adaptive threshold

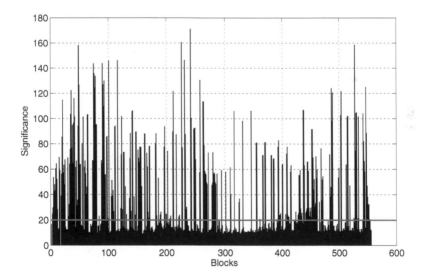

Figure 7.42 Significance values

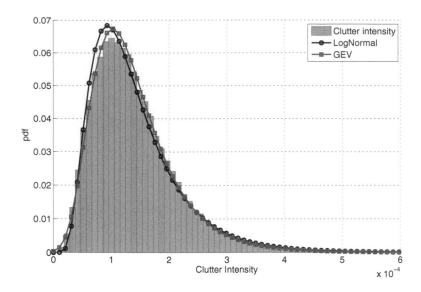

Figure 7.43 Probability density function of the sea clutter intensity computed from Figure 7.36

Figure 7.44 shows the results of the adaptive detector where targets are represented in black while the coast line is represented with a solid gray line.

In Figure 7.45 detection results have been superimposed to the SAR image after wavelet correlator, while Figure 7.46 shows a sub-image taken from Figure 7.45.

It is very interesting to note the detected targets near Storebaelt bridge. These targets are, presumably, cars crossing the bridge in both directions. Moving targets, in fact, appear both defocused and shifted in SAR images because their relative motion with respect to the SAR platform is different from that of the SAR scene center. In this case, the effect is that cars on the bridge appear in the SAR image displaced along the azimuth direction with respect to the bridge. They are easily recognized because they mainly stay along a straight line parallel to the bridge, indicating that many targets have the same velocity and direction. Some other cars appear in the other part of the bridge indicating cars traveling in the opposite direction.

7.3 ACKNOWLEDGMENTS

The authors thank the Italian Ministry of defence for partially funding this work under the framework of the "ISAR imaging of non-cooperative moving targets with Cosmo-SkyMed" (ISAR-SKY) project, and the Italian Space Agency (ASI) for providing COSMO Skymed spotlight SAR images.

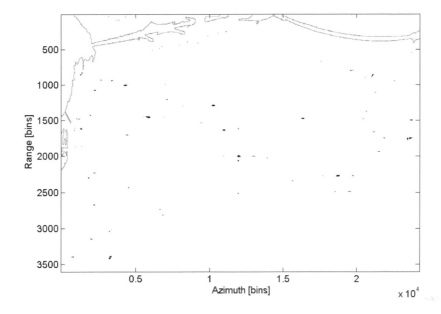

Figure 7.44 Detection results after post-processing

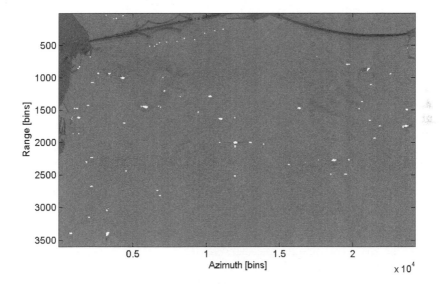

Figure 7.45 Detection results after post-processing

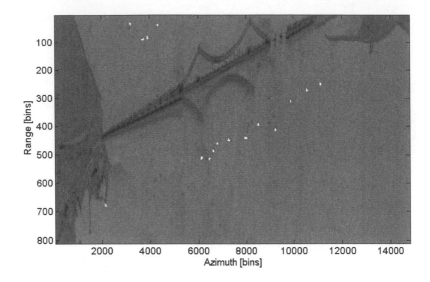

Figure 7.46 Detection results after post-processing

REFERENCES

1. Jiaqiu Ai, Xiangyang Qi, Weidong Yu, Yunkai Deng, Fan Liu, and Li Shi. A new cfar ship detection algorithm based on 2-d joint log-normal distribution in sar images. *Geoscience and Remote Sensing Letters, IEEE*, 7(4):806–810, Oct. 2010.
2. E.L. Christensen, N. Skou, J. Dall, K.W. Woelders, J.H. Jorgensen, J. Granholm, and S.N. Madsen. Emisar: an absolutely calibrated polarimetric l- and c-band sar. *Geoscience and Remote Sensing, IEEE Transactions on*, 36(6):1852–1865, Nov. 1998.
3. Ernesto Conte, M. Lops, and G. Ricci. Asymptotically optimum radar detection in compound-gaussian clutter. *Aerospace and Electronic Systems, IEEE Transactions on*, 31(2):617–625, April 1995.
4. D. J. Crisp. The state-of-the-art in ship detection in synthetic aperture radar imagery. *Report DSTO-RR-0272, Intelligence, Surveillance and Reconnaissance Division Information Sciences Laboratory*, May 2004.
5. M. Di Bisceglie and C. Galdi. Cfar detection of extended objects in high-resolution sar images. *Geoscience and Remote Sensing, IEEE Transactions on*, 43(4):833–843, April 2005.
6. R. A. Fisher and L. H. C. Tippett. Limiting forms of the frequency distribution of the largest or smallest memeber of a sample. *Mathematical proceedings of the Cambridge Philosophical Society*, 24(2):180–190, Oct. 2008.
7. F. Gini. Sub-optimum coherent radar detection in a mixture of k-distributed and gaussian clutter. *Radar, Sonar and Navigation, IEE Proceedings -*, 144(1):39–48, Feb. 1997.

8. B. Gnedenko. Sur la distribution limite du terme maximum d'une serie aleatorie. *Annals of Mathematics*, 44(3):423–453, 1943.

9. A. F. Jenkinson. The frequency distribution of the annual maximum (or minimum) values of meteorological elements. *Quarterly Journal of the Royal Meteorological Society*, 81(348):158–171, April 1955.

10. Jin Min Kuo and K.-S. Chen. The application of wavelets correlator for ship wake detection in sar images. *Geoscience and Remote Sensing, IEEE Transactions on*, 41(6):1506–1511, June 2003.

11. E. Makhoul, A. Broquetas, J.R. Rodon, Y. Zhan, and F. Ceba. A performance evaluation of sar-gmti missions for maritime applications. *Geoscience and Remote Sensing, IEEE Transactions on*, PP(99):1–14, 2014.

12. S.G. Mallat. A theory for multiresolution signal decomposition: the wavelet representation. *Pattern Analysis and Machine Intelligence, IEEE Transactions on*, 11(7):674–693, July 1989.

13. M. Martorella, D. Pastina, F. Berizzi, and P. Lombardo. Spaceborne radar imaging of maritime moving targets with the cosmo-skymed sar system. *Selected Topics in Applied Earth Observations and Remote Sensing, IEEE Journal of*, 7(7):2797–2810, July 2014.

14. R.L. Paes, J.A. Lorenzzetti, and D.F.M. Gherardi. Ship detection using terrasar-x images in the campos basin (brazil). *Geoscience and Remote Sensing Letters, IEEE*, 7(3):545–548, July 2010.

15. L.L. Scharf and D. Lytle. Signal detection in gaussian noise of unknown level: An invariance application. *Information Theory, IEEE Transactions on*, 17(4):404–411, July 1971.

16. D. Stagliano, A. Lupidi, and F. Berizzi. Ship detection from sar images based on cfar and wavelet transform. In *Advances in Radar and Remote Sensing (TyWRRS), 2012 Tyrrhenian Workshop on*, pages 53–58, Sept. 2012.

17. D. Stagliano, L. Musetti, D. Cataldo, A. Baruzzi, and M. Martorella. Fast detection of maritime targets in high resolution sar images. In *Radar Conference, 2014 IEEE*, pages 0522–0527, May 2014.

18. M. Tello, C. Lopez-Martinez, and J.J. Mallorqui. A novel algorithm for ship detection in sar imagery based on the wavelet transform. *Geoscience and Remote Sensing Letters, IEEE*, 2(2):201–205, April 2005.

19. M. Tello, C. Lopez-Martinez, and J.J. Mallorqui. Automatic vessel monitoring with single and multidimensional sar images in the wavelet domain. *ISPR Journal of Photogrammetry and Remote Sensing*, 61(3-4):260–278, Nov. 2006.

20. S. Wellek. *Testing Statistical Hypotheses of Equivalence and Noninferiority, 2nd ed.* Taylor & Francis, 2010.

8 Oil Spill Detection with SAR Images

F. Berizzi, A. Lupidi, and D. Staglianò

CONTENTS

Ocean pollution by oil slicks is one of the major environmental hazards. Oil tanker accidents (such as Exxon Valdez, Erika, and Prestige), although spectacular and highlighted, are responsible for only 5% of the total oil pollution worldwide, 95% coming from illegal discharges [4]. In the Mediterranean Sea, oil pollution monitoring is normally carried out by aircrafts or ships. This is expensive and is constrained by the limited availability of these resources. In

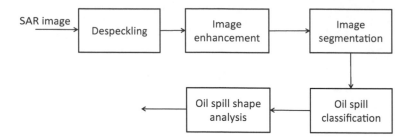

Figure 8.1   Oil-spill detection and classification algorithm

order to provide all-weather and global monitoring of such events, spaceborne synthetic aperture radar (SAR) has been recognized as a cornerstone. Radar imagery can provide a significant contribution for this field, identifying probable spills and tracking ships' routes over very large areas, then guiding aerial surveys for precise observations in specific locations. Automatic detection of oil spills in SAR images has been widely researched, with specific focus being given to the detection and classification of oil spills. In this chapter an algorithm for oil-spill detection and classification is proposed whose block scheme is shown in Figure 8.1 and whose main steps are

1. Despeckling, which aims at reducing the speckle noise on the SAR image for its better interpretation
2. Image enhancement
3. Image segmentation
4. Oil-spill classification
5. Oil-spill shape features extraction

A more detailed description of these steps is provided in the following sections. Results of the application of such algorithm to COSMO-SkyMed SAR images are shown in Section 8.6.

8.1   SPECKLE NOISE REDUCTION

An imaging radar generates a SAR image by transmitting a coherent electromagnetic wave and subsequently processing the backscattered signal from the illuminated objects. However, due to interference processes between scatterers, speckle noise is introduced into the image. Speckle noise is a disturbing factor because it limits the ability to correctly interpret SAR images, restricts edge extraction, image segmentation, target recognition and classification, and it introduces uncertainty in ground surface parametric inversion. Therefore, it is important to apply suitable speckle reduction methods prior to the image processing, which are able to smooth speckle noise, while retaining as much detailed information as possible. There are two types of speckle noise reduction techniques:

1. Multi-look processing, which involves the incoherent averaging of multiple looks of the same scene during the generation of the SAR image. This process narrows down the probability density function of the speckle noise and reduces the variance, but it also reduces the spatial resolution and improves the radiometric resolution. This simple technique is able to remove speckle noise efficiently, but much of the edge information is lost [3].

2. Spatial filtering, which is performed on the image domain and involves an analysis of the local statistics of each pixel.

While multilooking process is usually done during the data acquisition stage, speckle reduction by spatial filtering is performed on the image after it is acquired. No matter which method is used to reduce the effect of the speckle noise, the ideal speckle reduction method preserves radiometric information, the edges between different areas and the spatial signal variability (i.e., textural information). Based on the studies in [7], the Gamma-MAP filter was chosen. The filter has a 3×3 kernel size which is a good tradeoff between the noise reduction and the details preservation. The maximum a posteriori (MAP) filter is based on a multiplicative noise model with non-stationary mean and variance parameters. This filter assumes that the original digital number (DN) value lies between the DN of the pixel of interest and the average DN of the moving kernel. Moreover, many speckle reduction filters assume a Gaussian distribution for the speckle noise, but recent works have shown this to be an invalid assumption. In fact, natural areas have been shown to be more properly modeled as having a Gamma distributed cross-section. The Gamma-MAP algorithm incorporates this assumption. This model leads to the functional form of an optimum filter in the minimum mean square error (MMSE) sense for smoothing images. By using locally estimated parameter values, the filter is made adaptive so that the filter can be implemented in the spatial domain and is computationally efficient. The Gamma-MAP logic maximizes the a posteriori probability density function with respect to the original image. It combines both geometrical and statistical properties of the local area.

8.2 MORPHOLOGIC IMAGE ENHANCEMENT

This step deals with the improvement in the image quality. The morphological operation is performed by applying a morphological closing operator on the image to remove isolated and scattered pixels, or holes, in homogeneous areas, followed by the application of a morphological erosion operation to restore the previous boundaries. Closing, shown in Figure 8.2, is an important operator from the field of mathematical morphology. Like its dual operator, opening, it can be derived from the fundamental operations of erosion and dilation. These operators are normally applied to binary images, although there are gray level versions, as in SAR images. Closing is similar in some ways to dila-

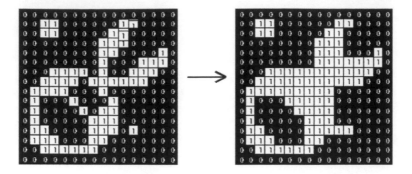

Figure 8.2 Closing operation (http://homepages.inf.ed.ac.uk/rbf/HIPR2)

tion in that it tends to enlarge the boundaries of foreground (bright) regions in an image and shrinks background color holes in such regions, but it is less destructive of the original boundary shape. As with other morphological operators, the exact operation is determined by a structuring element. The effect of the operator is to preserve background regions that have a similar shape to this structuring element, or that can completely contain the structuring element, while eliminating all other regions of background pixels. Closing is opening performed in reverse. It is defined simply as a dilation followed by an erosion using the same structuring element for both operations. The closing operator therefore requires two inputs: an image to be closed and a structuring element. Closing is the dual of opening, i.e., closing the foreground pixels with a particular structuring element is equivalent to closing the background with the same element.

Erosion, in Figure 8.3, is one of the two basic operators in the area of mathematical morphology, the other being dilation. The basic effect of the operator on a binary image is to erode away the boundaries of regions of foreground pixels (i.e., white pixels, typically). Thus, areas of foreground pixels shrink in size, and holes within those areas become larger. As for closing, the erosion operator takes two pieces of data as inputs. The first is the image which is to be eroded. The second is a (usually small) set of coordinate points known as a structuring element. It is this structuring element that determines the precise effect of the erosion on the input image.

As structuring element, in both cases, a 5 × 5 square mask was chosen to perform morphological enhancement. Formally, it is possible to write the following relation with the image Im_{in}:

$$Im_{out} = (Im_{in} \bullet E) \ominus E \tag{8.1}$$

where E is the structuring element, $\bullet$ is the closing operator and $\ominus$ is the erosion operator. The image is then filtered with a 5 × 5 Gaussian filter to eliminate the residual noise.

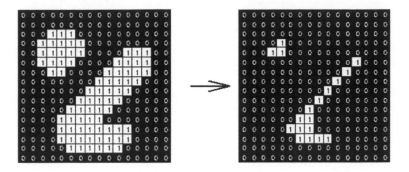

Figure 8.3   Erosion operation (http://homepages.inf.ed.ac.uk/rbf/HIPR2)

8.3   SEGMENTATION

Segmentation is the process that automatically selects group of pixels belonging to significant regions in the image, which can be labeled as belonging to a defined class. In this case, segmentation is necessary to automatically identify the areas with the lowest levels of gray, which may be associated with oil spills, as well as algae and areas of calm seas, e.g., areas where the wind is very weak. The fact that in these areas radar reflectivity is very low is due to the phenomenon of micro-surface wave damping caused by the oil or algae. In marine SAR images, in fact, the highest values of reflectivity are due to roughness caused by a more pronounced wave. Another problem is the image size. StripMap images are very large, and often automatic segmentation methods, which are typically based on a filtering operation, for example a wavelet filter, although valuable, need a considerable computational effort with often unacceptable long elaboration time. The segmentation algorithm proposed in [9] does not make use of any filtering operation thus resulting effective in terms of computation time. It consists in dividing the image into n_S sub-images, and estimating the histogram of the sub-block. This estimate was done with the kernel density estimation method. In statistics, kernel density estimation is a non-parametric way of estimating the probability density function (*pdf*) of a random variable. Kernel density estimation is a fundamental data smoothing problem where inferences about the population are made, based on a finite data sample. It is also known as the Parzen-Rosenblatt window method. Let $(p_1, p_2, \ldots, p_n)$ (where n indicates the number of pixels into a sub-image) be i.i.d. (independent and identically distributed) samples drawn from some distribution with an unknown density $f(p)$. Its kernel density estimator is

$$\tilde{f}_h(p) = \frac{1}{nh} \sum_{i=1}^{n} K\left(\frac{p - p_i}{h}\right) \tag{8.2}$$

where $K(\bullet)$ is the kernel, a symmetric, but not necessarily positive function that integrates to one, and $h > 0$ is a smoothing parameter called the bandwidth. A range of kernel functions is commonly used, namely, uniform, triangular, biweight, triweight, Epanechnikov, normal, and others. The Epanechnikov kernel is optimal in a minimum variance sense, though the loss of efficiency is small for the kernels listed previously, and due to its convenient mathematical properties, the "normal" kernel is often used. Based on these considerations, here a Gaussian kernel function is used. Ideally, the bandwidth should be as small as possible, but the variance of the estimator would be too high. On the other side, a large bandwidth cannot resolve a multimodal distribution. Value of h between 0.1 and 0.4 are usually chosen as a tradeoff between resolution and smoothness. By analyzing the various histograms, the threshold below which the pixels can be classified as a candidate oil spill can be chosen adaptively, as follows:

1. The algorithm first searches among all the local maxima of the *pdf* the smaller one, namely M_{min}, in other words the smallest local maximum.

$$M_{min} = \arg\min_{i}\{\mathbf{m}\} \qquad (8.3)$$

   where $\mathbf{m} = [m_1, m_2, \cdots, m_{n_S}]$ are the maxima of each *pdf*
2. Then, the algorithm searches for a threshold for each sub-images, namely λ_i where $i = 1, 2, \cdots, n_S$. When $f(p)$ exhibits a multi-modal behavior, as in Figure 8.4, the threshold coincides with the first local minimum. Such local minimum is found by calculating the first derivative of $f(p)$, namely $\dot{f}(p)$. When the *pdf* does not exhibit a multi-modal behavior, as in Figure 8.5 and Figure 8.6, the threshold in this case corresponds to the point whose amplitude deviates from the maximum of the *pdf* of a value which corresponds to the standard deviation toward the rightmost part of the histogram (brighter pixels)
3. Finally, among these candidates, sorted in ascending order, the optimum threshold, λ_{opt}, is chosen as the first value that has an amplitude greater than M_{min}

Figures 8.4 to 8.6 show the three cases that can be observed during the segmentation phase with the circles representing the respective tresholds.

It is very important that the distribution functions have a smooth behavior so as not to introduce false nulls in the derivative. Once the threshold is determined, all points with a lower value are classified as belonging to oil spill candidates.

The final step is the elimination of remaining isolated points or too small areas. To accomplish this, the algorithm analyzes the level of connectivity between the identified pixels; in other words, it defines which pixels are connected to other pixels. A set of pixels in a binary image that form a connected group is called an object or a connected component. The two-dimensional connectivities are

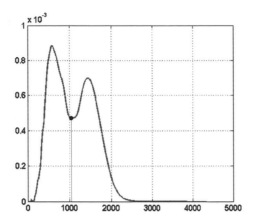

Figure 8.4 Double mode histogram

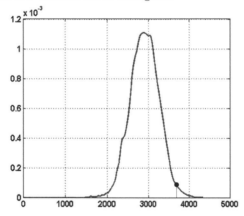

Figure 8.5 Sea histogram

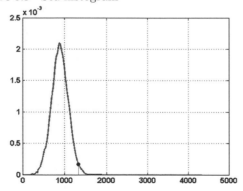

Figure 8.6 Spill or look-alike histogram

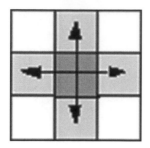

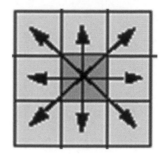

Figure 8.7   4-connected pixel

Figure 8.8   8-connected pixel

1. 4-connected (Figure 8.7): pixels that have exceeded the threshold are part of the same object when they are connected along the horizontal or vertical direction.
2. 8-connected (Figure 8.8): pixels that have exceeded the threshold are part of the same object when they are connected along the horizontal, vertical, or diagonal direction.

The connection algorithm is applied with a 3×3 square moving window over all the pixels of the image.

Then, the area of each object is calculated and objects with an area smaller than a pre-set threshold are discarded. For the purpose of oil-spill detection, the threshold can be set equal to 0.05 km^2. Then, objects whose area is smaller than 0.05 km^2 are discarded.

8.4   OIL SPILL CLASSIFICATION

This section deals with the description of the algorithm which extracts oil spill SAR signature. The algorithm is based on the assumption that the spatial spectral signature of sea areas has a fractal behavior (also named self-similar random process). It is possible then to differentiate dark areas on a few fractal descriptors. The first subsection describes mathematically the theory of self-similar processes highlighting two particular models, fractionally integrated autoregressive moving average (FARIMA) model and fractionally exponential (FEXP) model. The second subsection describes the method of estimation of radial power spectral density (PSD) and some fractal parameters into a reduced set of descriptors.

8.4.1   SELF-SIMILAR RANDOM PROCESS MODELS

As described in [1] oil on the sea surface dampens capillary waves, reduces Bragg's electromagnetic backscattering effect and, therefore, generates darker zones in the SAR image. A low surface wind speed, which reduces the amplitudes of all the wave components (not just capillary waves), and the presence of phytoplankton, algae, or natural films can also cause analogous effects (*look-alikes*). The analysis of natural clutter in high-resolution SAR images has been improved by the use of self-similar random process models. Moreover, many natural surfaces correspond to high-resolution SAR images that exhibit "long-term dependence" behavior and "scale-limited fractal" properties. In this section, by employing long-memory spectral analysis techniques, the discrimination between oil spills and low-wind areas in sea SAR images is described. Specifically, the long-memory or long-range dependence (LRD) property describes the high order correlation structure of a process. Suppose that $Y(x, y)$ is a discrete two-dimensional process whose samples are pixels. If $Y(x, y)$ exhibits long-memory, persistent spatial (linear) dependence exists even between distant observations. On the contrary, the short memory or short-range dependence (SRD) property describes the low order correlation structure of a process. If $Y(x, y)$ is a short-memory process, observations separated by a long spatial span are nearly independent. Among the possible self-similar models, two classes have been used to describe the spatial correlation properties of the scattering from sea: fractional Brownian motion (fBm) models and FARIMA models. In particular, fBm provides a mathematical framework for the description of scale-invariant random textures and amorphous clutter of natural settings. The fBm models allow for an efficient representation of stochastic self-similar processes of homogeneous and amorphous clutter, characterized by isotropic power spectral densities. The fBm model is a stochastic continuous and non-stationary process, for which the power spectral density cannot be calculated. It is possible anyway to provide a description of the process in power, applying a Fourier transform with random phase. The result is the representation of the model in function of a generalized PSD in the form:

$$E[S(F)] \propto F^{-\beta} \tag{8.4}$$

where F is the spatial frequency. If the fBm model provides a good fit with the periodogram of the data, it means that the PSD, as a function of the frequency, is approximately a straight line with negative slope (the β parameter) in a log-log plot. However, for a dataset in which the periodogram is characterized by different slopes in the log-log plane, the fBm model cannot be used to represent the PSD correctly. Therefore, FARIMA models that preserve the negative slope of the long-memory data PSD near the origin can be used, and through the so-called SRD functions, modify the shape and the slope of the PSD with increasing frequency. The SRD part of a FARIMA model is an auto regressive moving average (ARMA) process. A limitation to the applicability of FARIMA models is the high number of parameters required for the ARMA part of the PSD. Using an excessive number of parameters is undesirable because it increases the uncertainty of the statistical inference and the parameters become difficult to interpret. Then using FEXP models allows the representation of the logarithm of the SRD part of the long-memory PSD to be obtained, and greatly reduces the number of parameters to be estimated. FEXP models provide the same efficiency of fit as FARIMA models at lower computational costs. The first step in all the methods presented in this chapter is the calculation of the directional spectrum of a sea SAR image by using the 2-D periodogram of an N × N image. To decrease the variance of the spectral estimation, spectral estimates obtained from non-overlapping squared blocks of data are averaged. The characterization of isotropic or anisotropic 2-D random fields is done first using a rectangular to polar coordinates transformation of the 2-D PSD to obtain a directional spectrum, and then considering, as radial 1-D PSD, the average of the radial spectral densities. This estimated 1-D mean radial power spectral density (MRPSD) is finally modeled using a FARIMA or a FEXP model independently of the anisotropy of sea SAR images. In the next subsections an overview of the FARIMA and FEXP models (in the 1-D formulation) is presented.

8.4.1.1   Stationary Long-Range and Short-Range Dependence Processes

Let us suppose $X(1), X(2), \ldots, X(n)$ are random variables with mean value μ and variance σ^2 with the same marginal distribution D, sampled from the same process population. The autocovariance between $X(i)$ and $X(j)$ is defined as $\gamma(i,j) = E\{(X(i) - \mu)(X(j) - \mu)\}$ and the autocorrelation is $\rho(i,j) = \gamma(i,j)/\sigma^2$. For a stationary long-memory process the following properties hold:

1. The variance of the sample mean $var[\bar{X}(n)]$, with $\bar{X}(n) = E[X(n)]$ decays to zero at a slower rate than n^{-1} and is equal to a constant c_{var} times n^α for some $0 < \alpha < 1$.
2. The correlation $\rho(m)$ is asymptotically equal to a constant c_ρ times $|m|^{-\alpha}$ for some $0 < \alpha < 1$ (α same as point 1).
3. The spectral density $S(k)$ has a pole at zero that is equal to a constant c_f times $k^{-\alpha}$ for some $0 < \beta < 1$.
4. Near the origin, the logarithm of the periodogram $I(k)$ plotted versus the logarithm of the frequency appears to be randomly scattered around a straight line with negative slope.

Definition: Let $X(\nu)$, $\nu \in \mathbb{R}$ be a stationary process for which the following holds. If there are a real number $\alpha \in (0,1)$ and a constant $c_\rho > 0$ such that

$$\lim_{m \to +\infty} \frac{\rho(m)}{(c_\rho \cdot m^{-\alpha})} = 1 \qquad (8.5)$$

then $X(\nu)$ is called a stationary process with long memory or long-range dependence. On the contrary, stationary processes with summable correlations

$$\sum_{m=-\infty}^{+\infty} \rho(m) = constant < \infty \text{ and with exponentially decaying correlations}$$

$$\rho(m) \leq ba^m, \quad 0 < b < \infty, 0 < a < \infty \tag{8.6}$$

are called stationary processes with short memory or short-range dependence (SRD). Long-range dependence can also be defined by imposing a condition on the spectral density $S(k)$as follows.

Definition: Let $X(\nu)$, $\nu \in \mathbb{R}$ be a stationary process for which the following holds. There exists a real number $\beta \in (0,1)$ and a constant $c_f > 0$ such that

$$\lim_{k \to +\infty} \frac{S(k)}{\left(c_f \cdot |k|^{-\beta}\right)} = 1 \tag{8.7}$$

then $X(\nu)$ is called a stationary process with long memory or long-range dependence. With this definition $S(k)$ tends to infinity at the origin, i.e., $S(k)$ has a pole at zero.

8.4.1.2 Long-Range Self-Similar Processes

Let $Y(n)$, $n \in \mathbb{N}$ be a covariance stationary stochastic process with mean μ, variance σ^2, and autocorrelation function $\rho(m)$, $m \geq 0$. Assume that X has an autocorrelation function of the form $\rho(m) \approx c_\rho \cdot m^{-\alpha}$, as $m \to \infty$ where $0 < \alpha < 1$ and c_ρ is a constant. For each $l = 1, 2, \ldots, l \in (N)^+$ let $Y^{(l)}(n)$ denote the series obtained by averaging the original series $Y(n)$ over nonoverlapping blocks of size l

$$Y^{(l)}(n) = \frac{1}{l} \{Y[(n-1)l] + \cdots + Y(nl-1)\}, \quad n \geq 1. \tag{8.8}$$

Definition: A process is called (exactly) second-order self-similar with self-similarity parameter $H = 1 - \alpha/2$ if, for all $l = 1, 2, \ldots$, if

$$var[Y^{(l)}] = \sigma^2 l^{-\alpha} \tag{8.9}$$

and

$$\rho^{(l)}(m) = \rho(m) = \frac{1}{2}[(m+1)^{2H} - 2m^{2H} + |m-1|^{2H}], \quad m \geq 0 \tag{8.10}$$

where $\rho(l)$ denotes the autocorrelation function of $Y^{(l)}$.

Definition: A process is called (asimptotically) second-order self-similar with self-similarity parameter $H = 1 - \alpha/2$ if, for all m large enough so that $\rho^{(l)} \to \rho(m)$, as $l \to \infty$.

8.4.2 FARIMA MODEL DESCRIPTION

When dealing with LDR or SDR processes, FARIMA models can be used. FARIMA models are a natural extension of the classic ARIMA models. To simplify the notation, we consider only zero-mean stochastic processes. Let BK be a backshift operator defined so that the following equations hold:

$$Y(n) - Y(n-1) = (1 - BK)Y(n)$$
$$Y(n) - 2Y(n-1) + Y(n-2) = (1 - BK)^2 Y(n)$$
$$Y(n) - 3Y(n-1) + 3Y(n-2) - Y(n-3) = (1 - BK)^3 Y(n) \qquad (8.11)$$
$$\vdots$$

Let p and q be integers and define the polynomials:

$$\Phi(x) = 1 - \sum_{j=1}^{p} \Phi_j x^j \quad \Psi(x) = 1 - \sum_{j=1}^{q} \Psi_j x^j. \qquad (8.12)$$

Assume that all solutions of $\Phi(x) = 0$ and $\Psi(x) = 0$ are outside the unit circle. Let $W(n)$ for $n = 1, 2, \ldots, n \in (n)^+$, be i.i.d. normal variables with zero expectation and variance σ_w^2. An $ARIMA(p, q)$ model is defined to be the stationary solution of $\Phi[(BK)Y(n)] = \Psi[(BK)W(n)]$. If this last one holds for the dth difference:

$$\Phi[(BK)(1 - BK)^d Y(n)] = \Psi[(BK)W(n)] \qquad (8.13)$$

then $Y(n)$ is a $FARIMA(p, d, q)$ process for which d is an integer $d \geq 0$ and

$$(1 - BK)^d = \sum_{k=0}^{d} \binom{d}{k} - 1^k BK^k$$

Definition: Let $Y(n)$ be a discrete stationary process such that Equation (8.13) holds for some $|d| < 0.5$. Then $Y(n)$ is called $FARIMA(p, d, q)$ process.

Long-range dependence occurs for $0 \leq d < 0.5$ in 1-D process. The coefficient d is called "fractional differencing parameter", and for a fractal image, $0 \leq d < 1.5$ with the upper bound specific for a 2-D process.

8.4.3 FEXP MODEL DESCRIPTION

Definition: Let $r(k) : [-\pi, \pi] \to \mathbb{R}^+$ be an even function for which $\lim_{k \to 0} \frac{r(k)}{k} = 1$ and define $z_0 = 1$. Let $z_1, \ldots, z_q$ be smooth even functions in $[-\pi, \pi]$ such that the $n^* \times (q - 1)$ matrix A, where $n^* = (n-1)/2$, with column vectors $\left[z_l\left(\frac{2\pi}{n}\right), z_l\left(\frac{4\pi}{n}\right), \ldots, z_l\left(\frac{2n^*\pi}{n}\right) \right]^T$, $l = 1, \ldots, q$, is nonsingular for any n. Define a real vector $\boldsymbol{\eta} = [\eta_0, \eta_1, \ldots, \eta_q]$. A stationary discrete process $Y(m)$ with power spectral density function PSD given by

$$S(k, \boldsymbol{\eta}) = r(k)^d exp\left\{\sum_{l=0}^{q} \eta_l z_l(k)\right\},\tag{8.14}$$

with $0 \leq d < 0.5$, is defined as an FEXP process. The functions $z_1, \ldots, z_q$ are called short-memory components, whereas r is the long-memory component of the process. We can choose different sets of short-memory components, each one corresponding to a class of FEXP models. Two classes are particularly convenient because they are both characterized by the same LRD component given by

$$r(k) = |1 - e^{-ik}| = \left|2sin\left(\frac{k}{2}\right)\right|.\tag{8.15}$$

If the short-memory functions are defined as

$$z_l(k) = cos(lk), \quad l = 1, 2, \ldots, q\tag{8.16}$$

then the short-memory part of the PSD can be expressed as a Fourier series. If the short-memory components are equal to

$$z_l(k) = k^l, \quad l = 1, 2, \ldots, q\tag{8.17}$$

then the logarithm of the SRD component of the spectral density is assumed to be a finite order polynomial. In all that follows, this latter class of FEXP models is referred to as *polynomial FEXP models*.

8.4.4 SPECTRAL DENSITIES OF FARIMA AND FEXP PROCESSES

It is observed that a FARIMA process can be obtained by passing a fractional difference process through an ARMA filter. Therefore, in deriving the PSD of a FARIMA process, $S_{FARIMA}(k)$, it is possible to refer to the spectral density of an ARMA model, $S_{ARMA}(k)$, which is given by

$$S_{ARMA}(k) = \frac{\sigma_W^2 |\Psi(e^{ik})|^2}{2\pi |\Phi(e^{ik})|^2}.\tag{8.18}$$

The expression of $S_{FARIMA}(k)$ is

$$S_{FARIMA}(k) = |1 - e^{-ik}|^{-2d} S_{ARMA}(k) = \left|2sin\left(\frac{k}{2}\right)\right|^{-2d} S_{ARMA}(k).\tag{8.19}$$

The expression of the polynomial FEXP PSD can also be rewritten as follows:

$$S_{FEXP}(k, \boldsymbol{\eta}) = |1 - e^{-ik}|^{-2d} exp\left\{\sum_{l=0}^{q} \eta_l z_l(k)\right\} = \left|2sin\left(\frac{k}{2}\right)\right|^{-2d} S_{SRD}(k, \boldsymbol{\eta})$$

$$\tag{8.20}$$

where $S_{SRD}(k, \boldsymbol{\eta})$ denotes the SRD part of the PSD.

8.4.5   FRACTIONAL PARAMETERS METHOD DESCRIPTION

An image $Y(\mathbf{x})$, $\mathbf{x} \in \mathbb{R}^2$, of $n \times m$ pixels can be considered as a matrix with n rows and m columns whose elements are random variables. The autocorrelation of the image is defined as

$$R_{YY}(\mathbf{x}, \mathbf{y}) = E[Y(\mathbf{x}), Y(\mathbf{y})], \quad \mathbf{x}, \mathbf{y} \in \mathbb{R}^2. \tag{8.21}$$

If the autocorrelation function is invariant under all Euclidean motions, it depends only on the Euclidean distance between the points $\mathbf{x}$ and $\mathbf{y}$ and the field is referred to as homogeneous and isotropic $R_{YY}(\|\mathbf{u}\|^2) = E[Y(\mathbf{x}), Y(\mathbf{y} + \mathbf{u})]$. The PSD of the homogeneous field is a 2-D Fourier transform of $R_{YY}(u)$, $u = \|\mathbf{u}\|^2$. For sampled data, this is commonly implemented using the 2-D FFT algorithm. For a statistically isotropic and homogeneous field, it is a common practice to derive a 2-D model from a 1-D model by replacing $k \in \mathbb{R}$ in the PSD of a 1-D process, with $\|\mathbf{k}\|^2$, to get the radial PSD. On the contrary, the representation of non-isotropic fields requires additional information. Experimental results show that FARIMA and FEXP models always provide a good fit with the MRPSD of isotropic and non-isotropic sea SAR images. In particular, the proposed method utilizes a non-iterative technique to estimate the fractional differencing parameter d. To discriminate between low-wind and oil slick areas in sea SAR imagery two paramters can be used; the first is the LDR parameter d, which depends on the SAR image roughness, while the second is an SRD parameter which depends on the amplitude of the backscattered signal. Furthermore, it will be demonstrated that FEXP models allow for the computational efficiency to be maximized.

8.4.5.1   Estimation of Mean Radial PSD

In this method we refer to an $N \times N$ subimage that could correspond to a windy and clean sea area, to an oil slick (or spill) or to a low-wind area on the sea surface. To estimate the 2-D PSD $S(k_x, k_y)$ we use the 2-D periodogram $I(k_x, k_y)$ of the considered subimage. To decrease the variance of the spectral estimation, averaging of the spectral estimates over nonoverlapping squared blocks of data of size $n \times n$ (Bartletts periodogram) is performed. The samples of the resulting 2-D discrete periodogram are then represented as the elements of an $n \times n$ squared matrix. The transformation from Cartesian to polar coordinates returns a matrix whose elements are the samples of the 2-D discrete periodogram on a polar grid $I(k_{j,n}, \theta_i)$, which is the directional PSD, where $\theta_i = 2\pi i/n, i = 1, \ldots, n;$ $k_{j,n} = 2\pi j/n, j = 1, \ldots, n^*; n^* = \lfloor (n-1)/2 \rfloor$ where $\lfloor \cdot \rfloor$ represents the nearest lower integer mathematical operator.

The mean radial periodogram $I_{mr}(k_{j,n})$ is defined as

$$I_{mr}(k_{j,n}) = \frac{1}{n} \sum_{i=1}^{n} I(k_{j,n}, \theta_i). \tag{8.22}$$

8.4.5.2 Estimation of the Fractional Differencing Parameter

Least squares regression in the spectral domain is based on the analysis of the asymptotic behavior of the mean radial PSD at the origin. For the d parameter estimation, we apply linear regression to $log[I_{mr}(k_{j,n})]$ versus $log(k_{j,n})$ and calculate d from the slope of the resulting line. However, as the notion of long-memory is asymptotic and the spectral density of an LRD process is often proportional to k^{-2d} only in a small neighbourhood of zero, by wrongly assuming this proportionality for all the frequencies, the estimate of d can be highly biased. Consequently, an estimation based on all Fourier frequencies is of little practical importance, and we utilize the method only for a certain number of the smallest Fourier frequencies obtained by the search of the increment of the slope in the spectral domain. For doing that, the Savitzky-Golay filter is applied to the log-log MRPSD. The Savitzky-Golay method essentially performs a local polynomial regression (of degree l) on a series of values (of at least $t+1$ points which are treated as being equally spaced in the series) to determine the smoothed value for each point. The main advantage of this approach is that it tends to preserve features of the distribution such as relative maxima, minima and width, which are usually flattened by other adjacent averaging techniques. Then the Fourier frequencies are fitted to obtain a better precision in the searching of inflection points by discrete second derivative and we choose the nearest Fourier frequency to the inflection point for the d estimation. A two-step parameter estimation procedure can be used for the parameters of FARIMA processes.

From the expression

$$S_{FARIMA}(k_{j,n}) = \frac{S_{ARMA}(k_{j,n})}{\left|2sin\left(\dfrac{k}{2}\right)\right|^{2d}} \qquad (8.23)$$

assuming to know $I_{mr}(k_{j,n})$, an estimate of $S_{mr}(k_{j,n})$, with some algebra over m_R frequencies points, where m_R is the index denoting the last useful Fourier frequency, we can obtain the least squares estimator of d

$$\hat{d} = -\frac{\displaystyle\sum_{j=1}^{m_R}(u_j - \bar{u})(v_j - \bar{v})}{\displaystyle\sum_{j=1}^{m_R}(u_j - \bar{u})^2} \qquad (8.24)$$

where $v_j = log[I_{m_R}(k_{j,n})]$, $u_j = log\left[4sin^2\left(\dfrac{k_{j,n}}{2}\right)\right]$, $\bar{v} = \displaystyle\sum_{j=1}^{m_R}\dfrac{v_j}{m_R}$ and $\bar{u} = \displaystyle\sum_{j=1}^{m_R}\dfrac{u_j}{m_R}$.

8.4.5.3   Estimation of the ARMA Parameters and FARIMA Model

After determining the estimate $\hat{d}$ of the FD parameter d, the estimate of the SRD part of the mean radial PSD can be determined as

$$\hat{S}_{SRD}(k_{j,n}) = I_{m_R}(k_{j,n}) \left| 2sin\left(\frac{k}{2}\right) \right|^{2\hat{d}}, \tag{8.25}$$

which is the short-memory frequency response data vector defined as $\hat{h}(k_{j,n}) = \sqrt{\hat{S}_{SRD}(k_{j,n})}$. Then the parameters of a discrete transfer function that, for desired orders (n and m) of the numerator and denominator polynomials, corresponds to the frequency response are identified. The estimation of the real numerator and denominator coefficients of the transfer function in vector b and a is made by applying the classical Levi algorithm, whose output is used to initialize the Gaussian Newton method that directly minimizes the mean square error. The use of the damped Gauss-Newton method for iterative search guarantees the stability of the system. After choosing the maximum orders m_{MAX} and n_{MAX}, the estimation of vectors b and a for m and n with $m \in [0, m_{MAX}]$ and $n \in [0, n_{MAX}]$, respectively, are performed. The estimated transfer function is the one that corresponds to the estimated frequency response $h(k_{j,n})$ that gives the minimum root-mean-square error (RMSE) among all values of (m, n) with the $h(k_{j,n})$. The estimated ARMA part of the spectrum then becomes

$$\hat{S}_{ARMA}(k_{j,n}) = |\hat{h}(k_{j,n})|^2. \tag{8.26}$$

Then we calculate the FARIMA model PSD that fits the mean radial periodogram by multiplying the LRD component estimate, $\left| 2sin\left(\frac{k}{2}\right) \right|^{-2d}$, with the ARMA component estimate $\hat{S}_{ARMA}(k_{j,n})$. The best FARIMA model PSD approximation is the one that minimizes the RMSE between the data vector $I_{m_R}(k_{j,n})$ and the PSD of the FARIMA model

$$\hat{S}_{FARIMA}(k_{j,n}) = \frac{\hat{S}_{ARMA}(k_{j,n})}{\left| 2sin\left(\frac{k}{2}\right) \right|^{2\hat{d}}}. \tag{8.27}$$

8.4.5.4   Estimation of EXP Parameters and FEXP Model

Using FEXP models leads to the estimation of parameters in a generalized linear model. After determining $\hat{S}_{SRD}(k_{j,n})$ as in the method described above, the logarithm of the data $y_j = log[\hat{S}_{SRD}(k_{j,n})]$ is calculated and the vector $x_j = k_{j,n}$ is defined. Then the coefficients vector η of the polynomial $p(x)$ that fits the data $p(x_j)$ to y_j, in a least squares sense, are computed. The

SRD part of the FEXP model is equal to $\hat{S}_{EXP}(k_{j,n}) = exp\left\{\sum_{i=0}^{q}\eta(i)k_{j,n}^{i}\right\}$,

where q denotes the order of the polynomial $p(x)$. The estimated $\hat{S}_{EXP}(k_{j,n})$ is chosen as the one that gives the minimum RMSE with y_j among all values of q from q_{min} to q_{max} Then we calculate the FEXP model PSD that fits the mean radial periodogram by multiplying the LRD component estimate, $\left|2sin\left(\dfrac{k}{2}\right)\right|^{2\hat{d}}$, with the EXP component estimate $\hat{S}_{EXP}(k_{j,n})$.

8.4.5.5   Mean Amplitude of $\hat{S}_{SRD}(k_{j,n})$

The estimate of the self-similarity parameter H can be used as a measure of the roughness of high-resolution SAR images to discriminate between different clutter types. Unfortunately, with low surface wind speeds, experimental results obtainded show that the fractional differencing parameter, which depends on surface roughness in the spectral region of capillary waves (Bragg region) and represents the amplitude of the backscattered signal can be estimated with some difficulties. The mean amplitude of $\hat{S}_{SRD}(k_{j,n})$ is the short-range dependence parameter that is considered, defined as follows:

$$\bar{A}_{SRD} = \frac{1}{n^*}\sum_{j=1}^{n^*}\hat{S}_{SRD}(k_{j,n}). \tag{8.28}$$

Consider a SAR image of the sea surface with a region of lack of wind, and an oil slick. It was noted that too intense winds generate marine rough surfaces, but disperse the slicks very quickly and create a clutter (foam, spray), which makes it difficult to identify them on the SAR image. As an example a SAR image of the sea zone over which blows a wind of 10 m/s, shows dark areas that correspond to an oil slick (or spill) and a lack of wind. These two dark areas show very similar features and both are caused by the attenuation of the amplitudes of capillary waves that contribute to the Bragg resonance phenomenon. In the oil slick (or spill), the low amplitude of the backscattered signal is due to the concentration of the surfactant in the water that affects only the short wavelength waves, whereas in the low-wind area, the amplitudes of all the wind-generated waves are reduced (if the possible presence of the swell waves is excluded). In other words, the low-wind area tends to a flat sea surface, and it is expected that the spatial correlation in the correspondent portion of the SAR image decays to zero at a slower rate than that related to the oil slick (or spill) area. Since lower decaying correlations are related to higher values of the fractional differencing parameter d, it is foreseen that the low-wind area is characterized by the greatest d value in the whole image. Furthermore, given that the oil reduces only the amplitude of the capillary waves and does not affect all other wave components, the shapes of the mean radial PSD of oil slick (or spill) and clean water areas on the SAR image are expected to be very similar. This is because the oily substances that pollute the water cause only an attenuation of the amplitude of the backscattered signal (through the decrease of the Bragg resonance phenomenon), but, out of the Bragg wavelengths range, the same waves are present in the two areas. On the contrary, it is possible to guess that in the SAR image the shapes of the mean radial spectra should differ for low-wind and windy clean water areas. To understand what happens if the sea surface roughness decreases, it is assumed to examine a SAR image corresponding to a smooth sea area. It is possible to notice that lack of wind does not create capillary waves in the oil-free areas, making it impossible to detect the slicks in the SAR image. Therefore, the SAR image of a smooth sea surface, which clearly shows identifiable darker zones corresponding to a low-wind area and an oil slick, is considered. As the sea surface roughness decreases, only little differences between the considered SAR image anomalies are expected to estimate. Also the fractional differencing parameter should assume similar values for oil spills and low-wind areas. This is as a consequence of the sea state decreasing that, in the oil slick SAR image, leads to measure higher values of the spatial correlations as the lag tends to increase.

8.5 OIL SPILL SHAPE

After the segmentation and classification algorithms, a binary image of the oil spill is available. Also the gray level image of the only oil spill is available multiplying the binary one to the gray level scenario. After that, a cancellation of the small patches due to the speckle noise is applied, retrieving the concept of connectivity explained in Section 8.4. The next step is to automatically select only the patches that belong at the same oil spill. At this point, the binary image of the oil spill is composed by one or more patches that, since the oil spills are generated by moving ships, usually have a linear behavior and a specific direction. It is known that the detection of segments in digital pictures can be effectively performed by means of the Radon transform (RT), which concentrates the information about linear features of an image in a few high-valued coefficients (i.e., peaks) in the transformed domain [2] [6].

Using the RT and finding the peaks, it is possible to calculate the angle between the x-axis and the line in counterclockwise to have an estimation of the direction of the oil spill. First some spatial parameters of each dark patch are calculated, like [8] [5]:

Area: sum of the pixel belonging to the spot.
Perimeter: the distance between each adjoining pair of pixels around the border of the spot.
Major Axis Length: the length (in pixels) of the major axis of the ellipse that has the same normalized second central moments as the spot.
Minor Axis Length: the length (in pixels) of the minor axis of the ellipse that has the same normalized second central moments as the spot.

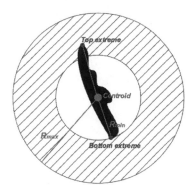 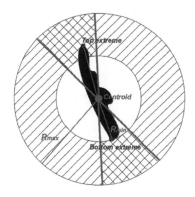

Figure 8.9   Spill circle delimitation and angular section

Max Intensity: the value of the pixel with the greatest intensity in the
   spot.
Mean Intensity: the mean of all the intensity values in the spot.
Min Intensity: the value of the pixel with the lowest intensity in the spot.
Orientation: the angle (in degrees, ranging from -90 to 90) by the RT.
Centroid: the center of mass of the spot.
Eccentricity: the ratio of the distance between the foci of the ellipse,
   that has the same second-moments as the region, and its major axis
   length. The value is between 0 for which the ellipse degenerates to a
   circle and 1, for which the ellipse degenerates to a line.
Complexity: describes how simple (or complex) the geometrical objects
   are. Here it is defined as $C = Perimeter/(2\sqrt{\pi \cdot Area})$ and for values
   near 1 it means that the spot is almost circular; for higher values the
   spot is more complex.
Extreme points in the region.

Then, to run the algorithm, the greatest dark patch is chosen and the
distances between its centroid and its extremes are calculated, as shown in
Figure 8.9. The reason why the greatest patch is selected is that it gives the
greater contribution for the RT and it might be enough to estimate the region
where to search a possible responsible ship. After that, the link of that patch to
another one is made looking for a pixel set to 1 in a region represented by a ring
centered in the centroid of the first patch with R_{MIN} equal to the minimum
distance between the centroid and the extremes and $R_{MAX} = R_{MIN} + 600$
pixels (value found experimentally).

Considering the linear behavior of the oil spill, it might be better to look
for another linked patch in the same direction of the greatest one. To do this,
an angular sector of 70° (value found experimentally), with its bisector in
the direction calculated by the RT, is intersected with the ring, as shown in
Figure 8.9. In this way an angular ring sector in which to look for another

linked patch is obtained. If some pixels set to 1 in the angular ring sector are found, the corresponding patch is linked to the first one and the algorithm focuses on the new linked patch and executes the same steps, and it continues until in the created angular ring sectors pixels set to 1 are no more found. The algorithm stops when in the angular sector there are no more pixels set to 1. At the end, the binary map of the oil spills is available.

8.6 DATA ANALYSIS

In this section, some results relative to the application of the proposed algorithm to images with oil spills and ships are presented. Specifically, two Cosmo-SkyMed SAR images of the area of Southern Adriatic Sea and Ionian Sea in front of Calabria, acquired respectively on August 22, 2009 (CSK-S2 mission) and on May 13, 2008 (CSK-S1 mission), both in HH polarization. In the images in Figure 8.10 and Figure 8.11, some dark spots (that will be classified) and several ships are evident. In particular, few scenarios containing dark areas and ships have been highlighted with rectangular boxes. Inside those, smaller rectangles identify the areas to be classified, which may be oil spills, sea or wind falls. Note that not all the considered subimages have the same size. Subimages of greater size provide lower variance of the 2-D periodogram. In the performed data processing, we have used

$$\begin{cases} m_{max} = 20 \\ n_{max} = 20 \\ q_{max} = 15 \end{cases}$$

for all the obtained results.

Figures 8.12 to 8.15 show the result of quality enhancement after morphological filtering for the Ionian Sea and Adriatic Sea dataset. The visual improvement is evident, and it is possible to see clearly the oil spill (elongated feature) and the low-wind area (in the upper-leftmost) part of the image of the Ionian dataset.

After the quality enhancement we show the results of the image segmentation and oil spills merging for the two datasets. For the Ionian dataset we can see that the algorithm (Figure 8.16 and 8.17), besides correctly revealing the whole oil spill, also detects two low-wind areas. The classification algorithm takes care of discriminating the type of zones. For the Adriatic sea dataset (see Figure 8.18 and 8.19), the algorithm has no problem in finding the oil slick, even if the largest is severely spread because of the possible age and the wind effects.

After the segmentation algorithm, the following five sub-images are analyzed to be classified. The two square sub-images numbered 1 and 2 in the Ionian Sea image (Figure 8.10) represent a clean sea area (1024 × 1024 pixels) and an oil spill (128 × 128 pixels), respectively. The three square sub-images, numbered 1, 2, and 3 in the Adriatic Sea dataset (Figure 8.11) represent a

Figure 8.10 Ionian Sea ROI

Figure 8.11 Southern Adriatic Sea ROI

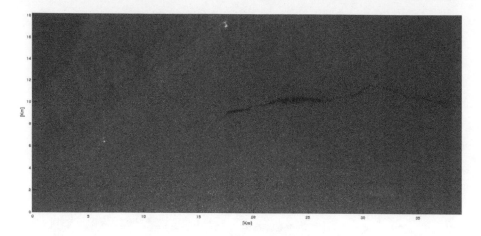

Figure 8.12 Ionian Sea original image

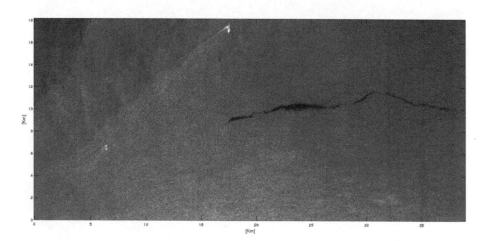

Figure 8.13 Ionian Sea after morphological enahancement

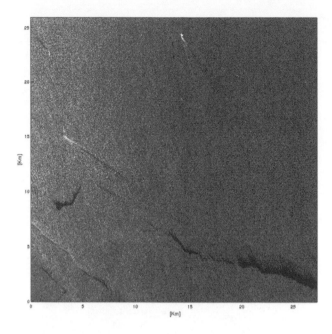

Figure 8.14 Adriatic Sea original image

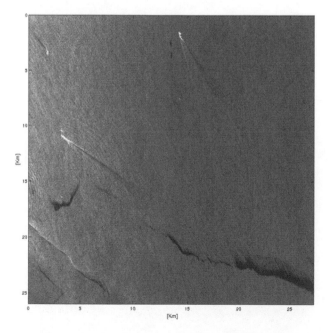

Figure 8.15 Adriatic Sea after morphological enahancement

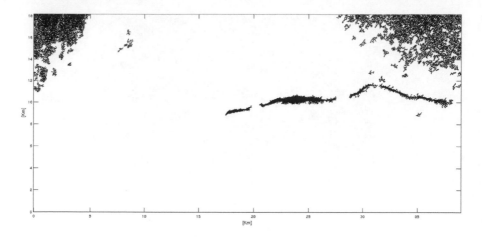

Figure 8.16 Ionian dataset image segmentation

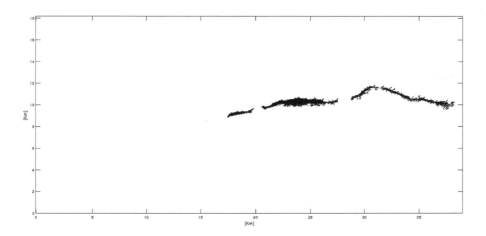

Figure 8.17 Ionian dataset image after spill classification

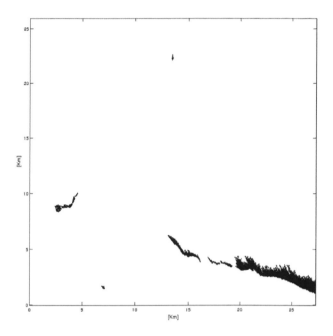

Figure 8.18 Adriatic dataset image segmentation

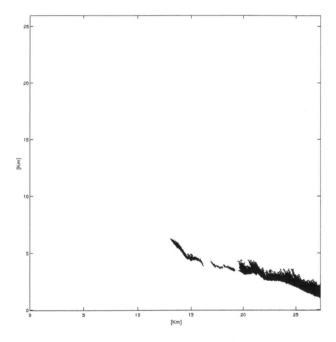

Figure 8.19 Adriatic dataset image after spill classification

Table 8.1

Ionian Sea satellite acquisition parameters

Ionian Sea	
Center latitude	38°15′27″ N
Center longitude	16°37′12″ E
Scene width	12673 pixels
Scene height	6433 pixels
Product start time (UTC)	13-MAY-2008 04:56:59.563026
Product end time (UTC)	13-MAY-2008 04:57:01.655571
Polarization	HH
Orbit	5038
PRF	3073.77 Hz
Radar Frequency	9600 MHz

Table 8.2

Southern Adriatic Sea satellite acquisition parameters

Southern Adriatic Sea	
Center latitude	40°40′54″ N
Center longitude	18°13′02″ E
Scene width	9896 pixels
Scene width	8211 pixels
Product start time (UTC)	22-AUG-2009 04:40:30.218055
Product end time (UTC)	22-AUG-2009 04:40:32.902178
Polarization	HH
Orbit	9215
PRF	3058.72 Hz
Radar Frequency	9600 MHz

clean sea area (1024×1024 pixels), a wind fall (512×512 pixels) and an oil spill (256×256 pixels), respectively. All the figures are represented with respect to the spatial frequency. Since the spatial resolution is 2.5 m for both the azimuth and the (ground) range coordinates, we have:

$$resolution = 2.5 \ [m]$$
$$ks_{j,n} = k_{j,n}/resolution \ [rad/m]$$

In the calculation of the periodogram, we averaged spectral estimates obtained from squared blocks of data containing 256×256 pixels (apart the oil

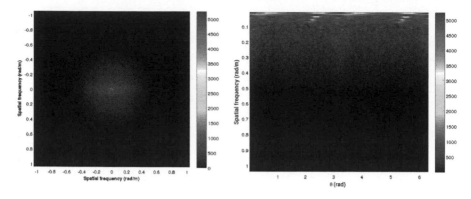

Figure 8.20 Periodogram (left) and polar periodogram (right) for the Adriatic Sea dataset

---

Table 8.3

Parameter estimates for the Ionian Sea dataset

	m_R	m	n	q	d	$\bar{A}_{SRD}$
Clean sea	4	14	19	15	0.1206	216.2745
Oil spill	3	1	17	15	0.1580	100.4706

---

spill of Figure 8.10 that contains 128×128 pixels). Figure 8.20 shows the periodogram and the polar periodogram of the image of (smooth) clean sea of the Southern Adriatic Sea. It is possible to define the sea as smooth because there are no visible peaks apart from the frequencies close to zero.

Figures 8.21 to 8.22 show the MRPSD in a log-log plot with FARIMA and FEXP models fitting for the clean sea subimage and the oil spill of the Ionian Sea dataset, respectively.

Figures 8.23 and 8.24 show the MRPSD in a log-log plot of the oil spill subimage with FARIMA and FEXP model fitting for the Southern Adriatic Sea dataset, while Figure 8.24 shows the fit of the wind fall subimage for the same dataset.

Tables 8.3 and 8.4 show the values of the estimated FARIMA and FEXP parameters m_R, m, n, q, d and $\bar{A}_{SRD}$ for both datasets.

From the values shown in Tables 8.3 and 8.4, the experimental results validate the proposed theory. The parameter $\bar{A}_{SRD}$ of the Adriatic Sea images presents low values and this is due to moderate wind velocities, but the discrimination between oil spill and low wind is very easy using either the d parameter or $\bar{A}_{SRD}$. In fact, the d parameter estimates corresponding to the

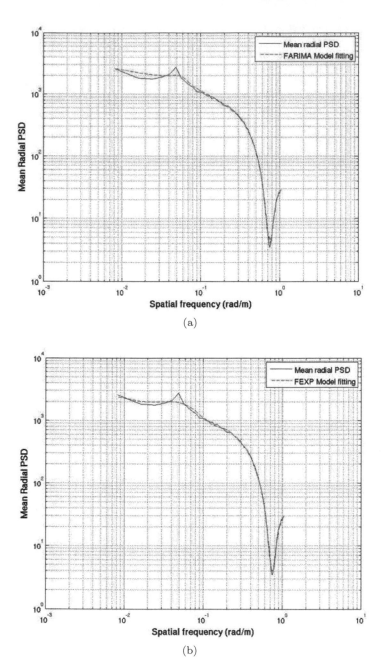

Figure 8.21 FARIMA (a) and FEXP (b) mean radial PSDs fitting for clean sea (Ionian Sea dataset)

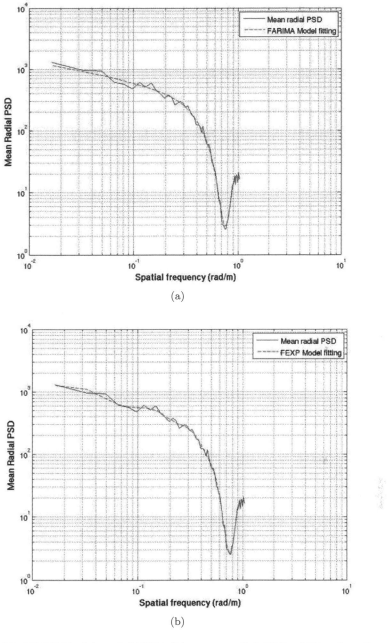

Figure 8.22 FARIMA (a) and FEXP (b) mean radial PSDs fitting for oil spill (Ionian Sea dataset)

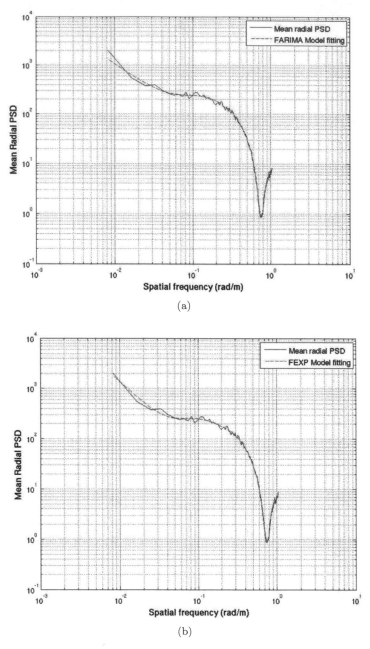

Figure 8.23 FARIMA (a) and FEXP (b) mean radial PSDs fitting for oil spill (Adriatic Sea dataset)

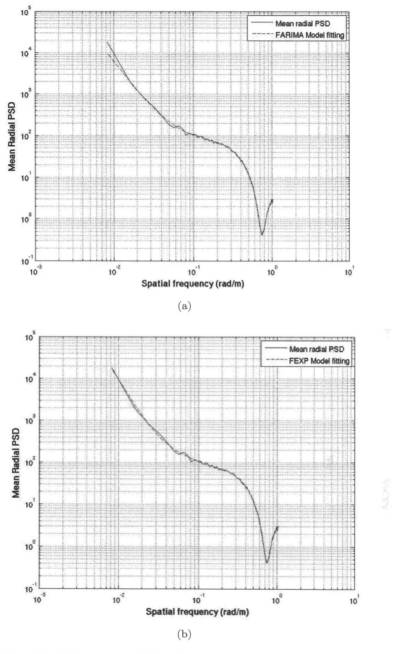

(a)

(b)

Figure 8.24 FARIMA (a) and FEXP (b) mean radial PSDs fitting for wind fall (Adriatic Sea dataset)

Table 8.4

Parameter estimates for the Southern Adriatic Sea dataset

	m_R	m	n	q	d	$\bar{A}_{SRD}$
Clean sea	5	14	14	15	0.3737	49.1312
Wind fall	7	16	13	15	1.2004	1.4208
Oil spill	5	18	16	15	0.5666	12.8820

oil spill and the clean water areas of Figure 8.11 are almost the same whereas the value corresponding to the wind fall is very different (it is higher). In other words, the low-wind area tends to a flat sea surface, and we expect that the spatial correlation in the corresponding portion of the SAR image decays to zero at a slower rate than that related to the oil spill area. This phenomenon is visible also in the Ionian Sea subimages; in fact, the values of the d parameter are very similar. For the $\bar{A}_{SRD}$, the amplitude of the signal backscattered from the sea surface, strongly reduced by the attenuation of the Bragg resonance phenomenon in the oil spill area, affects only the short-range dependence component whereas in the wind fall area all the wave components are reduced and then the SRD parameter is lower than in the oil spill and clean sea.

After the classification algorithm, the linear direction of the oil spill and several spatial features of the two scenarios have been analyzed. The RT is applied to the binary images of Figures 8.17 and 8.19, obtained after the oil spill merging algorithm, to estimate the linear direction of the oil spill, and the results are shown in Figure 8.25 for the Ionian and Adriatic datasets. On the y-axis and x-axis, the distance ρ in pixel from the center of the image under test and the angle θ in degrees ($°$) between the x-axis and the straight line, which estimates the linear behavior of the oil spill, are represented, respectively.

The angles of the straight lines and the spatial features extracted are shown in Tables 8.5 and 8.6.

As a final example the processing on a real validated case is presented. On August 14, 2011 a leak in a flow line leading to the Gannet Alpha oil platform 180 km off Aberdeen was found by Marine Scotland. A Cosmo-SkyMed image of this scenario with 3-m resolution was acquired on August 19, 2011. The validation of the oil spill was performed by the Italian company e-GEOS. The characteristics of the dataset are reported in Table 8.7

Figure 8.26 and 8.27 show the dataset before and after the image enhancement algorithm, respectively. Contours of the slick are cleary visible, where also some features regarding the direction of the sea waves are visible.

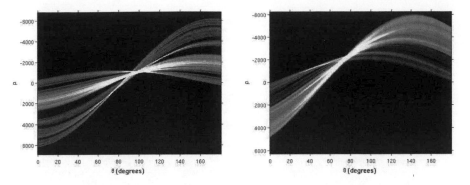

Figure 8.25   Estimation of spill direction angle after spill merging, Ionian Sea dataset (left) and Adriatic Sea dataset (right)

Table 8.5
Slick characteristics, Ionian Sea dataset

Ionian Sea

Angle straight line	$10°$
Area	460919 pixels
Perimeter	$2.532 \cdot 10^4$ pixels
Complexity	8.5312
Major Axis Length	$6.9053 \cdot 10^4$ pixels
Minor Axis Length	523.4341 pixels
Eccentricity	0.9971

Table 8.6
Slick characteristics, Adriatic Sea dataset

Southern Adriatic Sea

Angle straight line	$-11°$
Area	1026311 pixels
Perimeter	$2.4961 \cdot 10^4$ pixels
Complexity	8.5312
Major Axis Length	$6.9506 \cdot 10^3$ pixels
Minor Axis Length	523.4341 pixels
Eccentricity	0.9955

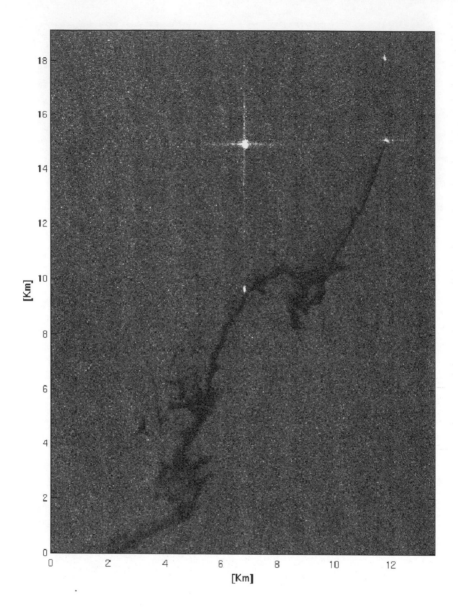

Figure 8.26 Northern Sea Platform dataset, original image

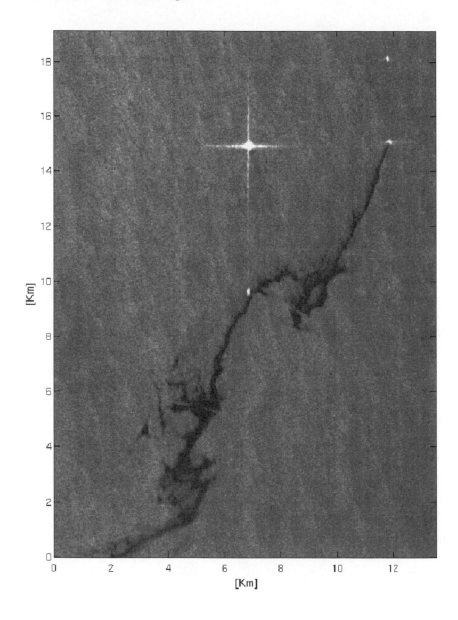

Figure 8.27 Northern Sea Platform dataset, after morphological enhancement

Table 8.7

Northern Sea satellite acquisition parameters

Southern Adriatic Sea

Center latitude	57°09′31″ N
Center longitude	0°47′06″ E
Scene width	20544 pixels
Scene width	20678 pixels
Product start time (UTC)	19-AUG-2011 05:31:35.3976261
Product end time (UTC)	19-AUG-2011 05:31:42.7719585
Polarization	VV
PRF	3065.04 Hz
Radar Frequency	9600 MHz

Table 8.8

Parameter estimates for the Northern Sea dataset

	d	$\bar{A}_{SRD}$
Clean sea	0.3901	64.2173
Oil spill	0.3906	39.6219

In these images there are no look-alikes (low wind areas or algae patches), so in this case the segmentation algorithm and the classification algorithm have the same result, shown in Figure 8.28.

Figure 8.29 shows the periodogram and the radial periodogram for the Northern Sea dataset, utilized to estimate the mean PSDs shown in Figure 8.30, where the PSD of the oil spill with FARIMA and FEXP method, respectively, are shown.

In Table 8.8 the parameters d and $\bar{A}_{SRD}$ for this dataset are shown, highlighting a consistency with the classification parameters.

8.7 ACKNOWLEDGMENTS

The authors thank the Italian Space Agency (ASI) for partially supporting and funding this work under the framework of the first announcement of opportunity for the exploitation of COSMO Skymed products, and for providing COSMO Skymed spotlight SAR images.

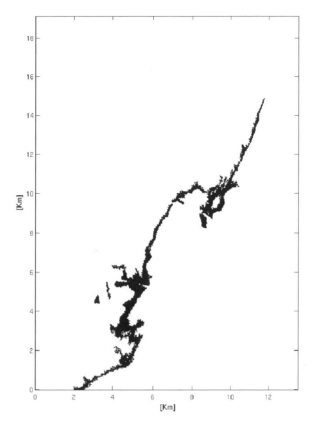

Figure 8.28   Northern Sea platform dataset, spill extraction via segmentation algorithm

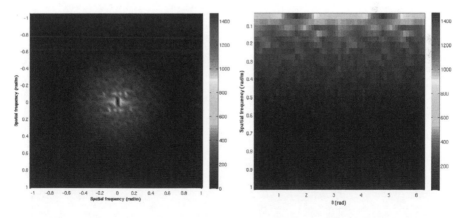

Figure 8.29   Periodogram (left) and polar periodogram (right) for the Northern Sea dataset

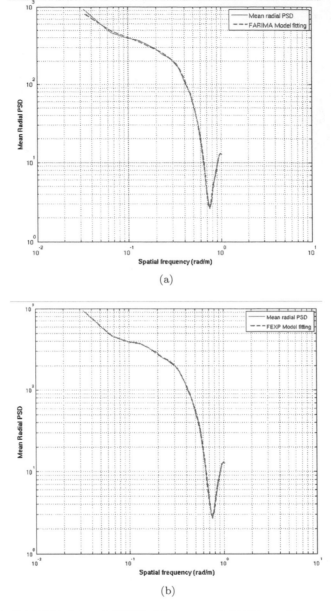

Figure 8.30 FARIMA (a) and FEXP (b) mean radial PSDs and fitting for oil spill (Northern Sea dataset)

REFERENCES

1. M. Bertacca, F. Berizzi, and E. Dalle Mese. Long-range dependence models for the analysis and discrimination of sea-surface in sea sar imagery. *Image Processing for Remote Sensing*, (2):189–223, 2007.
2. Bracewell and Ronald N. *Two-Dimensional Imaging*. Prentice Hall, Englewood Cliffs, NJ, 1995.
3. V. S. Frost, J. A. Stiles, K. S. Shanmugan, and J. C. Holtzman. A model for radar images and its application to adaptive digital filtering of multiplicative noise. *IEEE Transactions on Pattern Analysis and Machine Intelligence*, PAMI-4(2):157–166, March 1982.
4. F. Galland, P. Refregier, and O. Germain. Synthetic aperture radar oil spill segmentation by stochastic complexity minimization. *IEEE Geoscience and Remote Sensing Letters*, 1(4):295–299, Oct. 2004.
5. A. Gambardella, G. Giacinto, and M. Migliaccio. On the mathematical formulation of the SAR oil-spill observation problem. In *Geoscience and Remote Sensing Symposium, 2008. IGARSS 2008. IEEE International*, volume 3, pages III – 1382–III – 1385, July 2008.
6. Jae S. Lim. *Two-Dimensional Signal and Image Processing*. Prentice Hall, Englewood Cliffs, NJ, 1990.
7. M. Mansourpour, M. A. Rajabi, and J. A. R. Blais. Effects and performance of speckle noise reduction filters on active radar and SAR images. *ISPRS*, 2007.
8. A. H. S. Solberg, C. Brekke, and P. O. Husy. Oil spill detection in radarsat and envisat sar images. *IEEE Trans. Geosci. Remote Sens.*, 45(3), 2007.
9. D. Stagliano, A. Lupidi, F. Berizzi, and M. Martorella. Exploitation of cosmo-skymed system for detection of ships responsible for oil spills. In *Geoscience and Remote Sensing Symposium (IGARSS), 2012 IEEE International*, pages 915–918, July 2012.

9 Non-Cooperative Moving Target Imaging

M. Martorella, F. Berizzi, E. Giusti, and A. Bacci

CONTENTS

Synthetic aperture radar (SAR) systems were originally employed in Earth observation applications, such as ocean, land, ice, snow and vegetation monitoring, among others [17], [5], [2]. Nevertheless, the ability to form high resolution images from remote platforms in all day/all weather conditions has also rapidly made SAR systems became very attractive for military and homeland security applications.

In such applications, man-made non-cooperative target imaging becomes the main interest rather than the observation of natural phenomena.

Nevertheless, many SAR processors, which are designed to form highly focused images at very high resolution, are based on the assumption that the illuminated area is static during the synthetic aperture formation [2]. As a consequence, standard SAR techniques are typically unable to focus moving targets while forming a focused image of the static scene, leading to blurred and displaced images of any object that is not static during the synthetic aperture formation.

Many papers in literature deal with the problem of moving targets motion compensation. In [8] an algorithm for moving targets detection in SAR images and estimation of their motions parameters is proposed. In the same paper, however, the authors compensate for the Doppler shift by simply inserting the targets at the estimated position, and without performing phase compensation to eliminate motion-induced phase errors. Moreover, the algorithm requires a sequence of SAR images of the same target to properly estimate its motion parameters.

In [15],[22], [21], [6], the authors attempt to compensate the relative motion between the SAR antenna phase center and the moving targets by first estimating the target motion parameters and then by using the estimates in phase compensation. These papers, although proposing different methods, are based on the assumption that targets move along rectilinear trajectory with uniformly accelerated motion. In a real scenario, however, air and maritime targets undergo translational and angular motions due, for example, to the maneuverings or the sea state so that a more proper target motion model should be accounted for.

In [19], [20] more complex target motions are considered. Specifically, both the papers aim at compensating the translational and rotational target motion. These methods, however, require the presence of multiple prominent scattering centers on the target that should be tracked across the aperture. In order to properly estimate the translational and rotational motions, the authors state that at least 4 scatterers are required for each target. It could be difficult, however, to accurately isolate the received signal of such scatterers because the target image is defocused and because the target is usually superimpose upon a clutter background.

Inverse synthetic aperture radar (ISAR) has been suggested as an alternative way way to look at the problem of forming a synthetic aperture to obtain high resolution images of non-cooperative targets [1],[18]. ISAR techniques do not base their functioning on the assumption that the target is static during the synthetic aperture formation. Conversely, the target motion is crucial for synthetic aperture formation. ISAR techniques do not require the knowledge of a priori information about the target motion, but some other issues must be taken into account in ISAR image formation process. Typical problems are related to the cross-range scaling, the identification of the image plane and the fact that the imaging system performance is not entirely predictable. Other image formation constraints may include the image size and the achievable

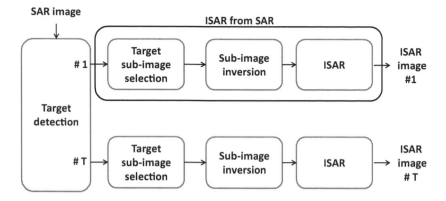

Figure 9.1 Block scheme of the detection and refocusing processor

cross-range resolution [11, 10, 4]. Notwithstanding, ISAR imaging provides acceptable solutions when SAR imaging fails, as it will be proven in the following sections.

9.1   ISAR FROM SAR PROCESSING CHAIN

A solution based on the application of the ISAR processing to compensate the unknown part of the relative motion between radar and non-cooperative moving target is proposed. The functional block diagram is shown in Figure 9.1 and it is composed of the following main steps:

1. Target Detection. Every target to be refocused must be detected within the SAR image first. This is a crucial step because the unknown target motion leads to dispersion of the useful target energy over a wide region. Detection is critical when dealing with ground target because of strong clutter.
2. Target sub-image selection. Once detected, every target must be cropped and separated from every contribution due to clutter and other targets. In this way, a number of sub-images equal to the number of detected targets is obtained.
3. Sub-image inversion. Since the ISAR processing takes raw data as input, an inversion mapping from the target image to the target raw data is mandatory. It is worth highlighting that raw data obtained from the whole SAR image is not a suitable input for the ISAR processing because it usually contains the returns from several moving targets, each one with its own motion. ISAR processing must be applied to a single target raw data at each time.
4. ISAR processing. It performs motion compensation and image formation.

Details about land masking and target detection are given in Chapter 7. In the next sections a method to discriminate moving and stationary targets will be introduced. Then, details about crop selection and sub-image inversion steps will be given. Finally, an extensive result analysis will be performed.

9.2   MOVING TARGETS DETECTION

In this section, a method to discriminate whether a detected target is a moving or stationary will be introduced. This is an important task for two reasons. First, the information about the target motion can be of interest in a surveillance scenario and second, if the detected target is stationary some operations, such as the cross-range scaling, can be avoided leading to a reduction in the processing time. In fact, in case of stationary targets, the cross-range axis is the same as the original SAR image azimuth axis.

A test to automatically discriminate moving from stationary targets can be formulated and implemented. The test consists of the evaluation of a decision parameter (PAR) which is then compared to threshold.

The measurement of the similarity is performed by evaluating the cross-correlation function between the image of the target before and after the autofocus process. This cross-correlation function is then compared to the autocorrelation function of the image before autofocus. Since, in case of stationary target the images before and after autofocus will be very similar, the cross-correlation function and the auto-correlation function will be very similar too, presenting a high and narrow peak. Conversely, in the case of a moving target, the cross-correlation function will present a lower peak and a smoothed shape.

Since the auto-correlation function and the cross-correlation function present this particular shape, the percentage difference between the image contrast (IC) (defined in Chapter 3) of the autocorrelation and the cross-correlation functions can be compared to a threshold to detect if the target is stationary or moving.

The images of a stationary target before (a) and after (b) the autofocus process are shown in Figure 9.2. The target is stationary because, as it is evident from a visual point of view, the autofocus process does not lead to any improvement in the image quality. As stated before, the autocorrelation function, Figure 9.2 (c), and the cross-correlation function, Figure 9.2 (d), are very similar with a high peak. The image contrast values are $IC_{auto} = 1.3036$ and $IC_{cross} = 1.3034$ with a difference of 0.019%

The images relative to a moving target before (a) and after (b) the autofocusing process are shown in Figure 9.3.

In this case the improvement in the image focus due to the autofocus process is very evident: This may be noted as the cross-correlation function appears much smoother than the autocorrelation function. In this case, the IC values are $IC_{auto} = 1.3715$ and $IC_{cross} = 1.3361$ with a difference of 2.58%

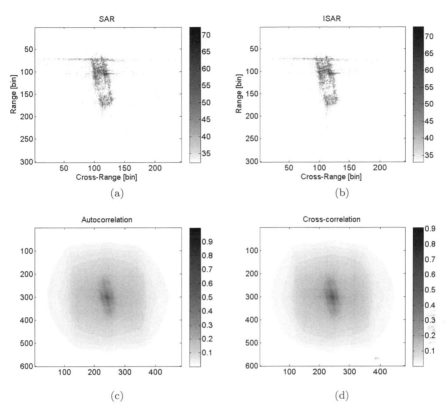

Figure 9.2 Ship 1 SAR (a), ISAR (b), autocorrelation function (c), cross-correlation function (d)

A threshold on the percentage variation (PAR) is heuristically set to 0.5% after investigating several cases.

9.3 TIME-WINDOWING ENABLING

Once a moving target has been detected, it could be useful to automatically detect whether the time-windowing algorithm is needed to form well-focused ISAR images of the target. The time-windowing algorithm is usually needed when a target is maneuvering and then experiences complex motions. In this event, the hypothesis of constant target effective rotation vector usually does not hold. As a consequence, the integration time must be shortened to fulfil such a requirement. The central time instant and the time length of the time window must be suitably chosen and this is the purpose of the time-windowing algorithm. The proposed algorithm is based on the cross-correlation between two sub-apertures of the same data. The idea behind this algorithm is that the

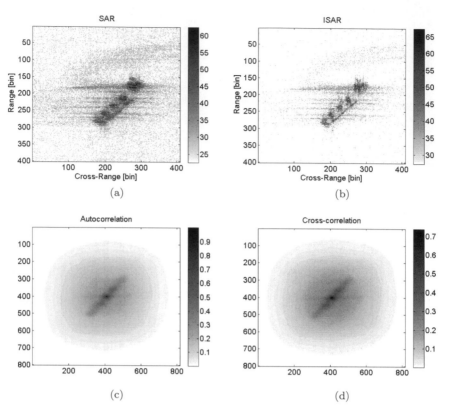

Figure 9.3 Ship 2 SAR (a), ISAR (b), autocorrelation function (c), cross-correlation function (d)

two ISAR images corresponding to the two sub-apertures should be different when the target experiences complex motions.

The algorithm can be summarized as follows:

1. The autofocusing process is applied to the whole observation time.
2. Two sub-apertures data are focused by exploiting the target motion parameters estimated at point 1 and two ISAR images are then obtained, namely I_{f_1} and I_{f_2}.
3. The cross-correlation between I_{f_1} and I_{f_2} is computed and its peak compared to a given threshold.

If the two ISAR images obtained by using the same target motion parameters are very similar, then the time-windowing algorithm is not needed. The similarity between the two ISAR images is measured via the cross-correlation peak. A threshold has been set heuristically at 0.7. The time-windowing algorithm is applied when the cross-correlation peak is lower than this threshold.

In Figure 9.4 a case in which the time-windowing algorithm is needed is shown. In this case, the target under test is a small manoeuvring target and the refocusing process applied to the whole observation time cannot produce a well-focused image as shown in Figure 9.4 (b). The two ISAR images obtained by processing the two sub-aperture of the same data are instead shown in Figure 9.4 (c) and (d). Finally, the cross-correlation between them is shown in Figure 9.4 (e). The peak of the cross-correlation is in this case equal to 0.63, which is lower than the threshold indicating that the time-windowing is necessary. In fact, as can be noted from a visual inspection of the two ISAR images, the target experiences complex motions as it changes orientation within the whole observation time.

9.4   SUB-IMAGE INVERSION

After detection and crop extraction a set of sub-images containing each single target is available. Since the natural input of ISAR processing is the raw data an inversion step is needed to go from the image domain to the data domain. This operation depends on the SAR image formation algorithm used to obtain the whole SAR image. It is worth highlighting that, even if available, the SAR raw data cannot be used as input of the ISAR processing because this data contains the superimposition of the signal received by the static scene and all the moving targets and each components have a different motion to be compensated.

In the next subsection the inversion of some SAR image formation algorithms will be investigated and some conclusion will be raised. Specifically the inversion of the range Doppler, the polar format [3], the $\Omega - k$ and the chirp scaling [2, 7] will be treated.

9.4.1   INVERSE RANGE DOPPLER

As stated in Chapter 3 when the total aspect angle variation is not too large and when the effective rotation vector is sufficiently constant during the observation time, the range Doppler technique represents an accurate and computationally effective tool for SAR/ISAR image reconstruction. Under these constraints, the spatial frequency domain in which the received signal is defined can be assumed to be a nearly regularly sampled rectangular grid. Therefore, a two-dimensional fast Fourier transform (FFT) can be used to reconstruct the image. In this case the inversion algorithm, namely inverse range Doppler (IRD), consists of a Fourier inversion, which is usually implemented via a two-dimensional inverse fast Fourier transform (IFFT). Moreover, it is worth pointing out that the inversion involves a new spatial frequency domain that must be defined for each crop. As the resolution of the whole SAR image and the resolution of each crop are equal, the observation time and the bandwidth of the data obtained via the inversion are the same as the SAR raw data. On the other hand, since the crop is only a portion of the whole SAR image

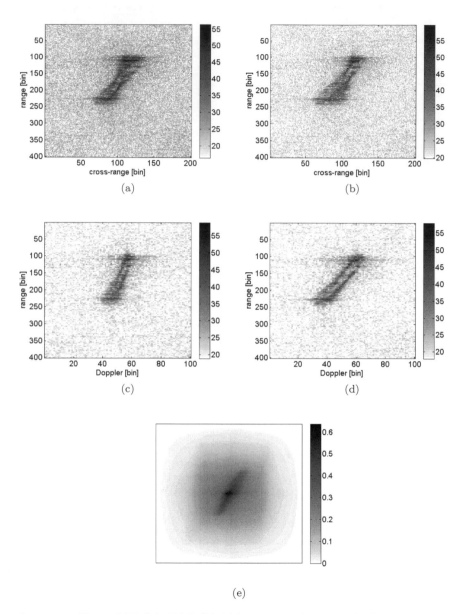

Figure 9.4 Ship 2 SAR (a), ISAR (b), ISAR image relative to the first sub-aperture (c), ISAR image relative to the second sub-aperture (d), and cross-correlation function (e)

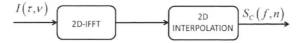

Figure 9.5   Inverse polar format block diagram

Figure 9.6   Inverse $\omega - k$ block diagram

the PRF and the frequency sample spacing in the equivalent raw data can be evaluated as

$$PRF = \frac{N_{crop}}{NT_R}$$
$$\Delta f = \frac{B}{M_{crop}} \tag{9.1}$$

where N_{crop} and M_{crop} are the crop samples in the cross-range and range dimension, respectively.

9.4.2   INVERSE POLAR FORMAT

A description of the polar format algorithm can be found in [2, 3]. The inverse polar format (IPF) can be obtained by tracing back the steps used to form the image. Specifically, a 2D-IFFT is applied to the SAR sub-image followed by a 2D inverse interpolation and a spatial frequency domain mapping. The functional block of the IPF is shown in Figure 9.5.

Unfortunately, the interpolation is not invertible leading the introduction of some error in attempting any type of inversion.

9.4.3   INVERSE $\omega - K$

The $\omega - k$ algorithm is described in [2, 5] and, as the PF, makes use of an interpolation that makes this algorithm not exactly invertible. The inverse $\omega - k$ (IOK) algorithm proposed in this work has been obtained by inverting each single step of the direct algorithm, as shown in Figure 9.6. As all steps in the direct algorithm are invertible except for the interpolator, we will expect that artefacts may be introduced by the interpolation inversion, which is implemented by means of an additional interpolation.

9.4.4   INVERSE CHIRP SCALING

Differently from the PF and the $\omega - k$, the CS algorithm does not need an interpolation step, therefore resulting in a more computationally efficient algorithm. As stated in [16, 13, 14], the CS algorithm addresses the problem of equalizing the range migration of all the point scatterers composing the

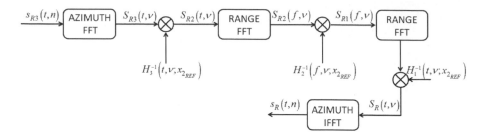

Figure 9.7 Inverse chirp scaling block diagram

target. Since all the scatterers follow the same trajectory, they can be compensated by a known phase term. The inverse chirp scaling (ICS) algorithm flow chart is shown in Figure 9.7, in which the steps involved in the chirp scaling algorithm are reversed.

9.5 ISAR FROM SAR: RESULTS

In this section results of the proposed techniques are shown which refer to different radar imaging systems both airborne and spaceborne. The whole ISAR processing chain which is composed of target motion compensation, cross-range scaling and when necessary the time-widowing approach has been applied.

Before going through the results a comment should be made concerning the inversion step. In [12] and [9] a comparison between the IOK and the IRD has been made. More specifically, CSK SAR images formed by using the OK algorithm have been processed. The aim of that study was to assess whether the IRD can be used even when the SAR images are formed with different SAR imaging algorithm. In [12] and [9], ISAR images of some targets have been obtained by using both the IOK and the IRD algorithm. The results, in terms of IC, prove that at least for small scenes both the inversion methods are effective. It should be said that even if appreciable differences are not evident between the two methods, both the IRD and the IOK may introduce phase errors. The former because it is an approximation way of inverting the SAR image and the latter because of the interpolation step, which is not invertible. This holds also for IPF but not for the ICS which is invertible as it does not require an interpolation step.

Phase errors eventually introduced via the inversion may affect the useful information for phase-related post-processing algorithm such that interferometry.

In what follows, the IRD algorithm will be used for the inversion as it is the easiest and fastest method.

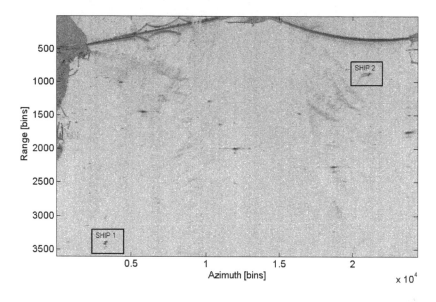

Figure 9.8 SAR image with ships highlighted

9.5.1 EMISAR DATASET

In this sub-section a SAR image acquired by means of the EMISAR system is considered. Figure 9.8 shows the SAR image where the ships processed in this sub-section have been circled. Parameters concerning this SAR image can be found in Table 7.3.

Both the targets have been first processed by the *"moving target detector"* to assess if they are moving targets. Once the targets have been recognized as moving targets the other detector assessing whether the time-windowing algorithm is needed is then applied. For both these two targets the time-windowing is unnecessary, then the target motion compensation and the cross-range scaling are applied to get an ISAR image in a fully spatial domain, namely the range/cross-range domain.

Figure 9.9 shows the sub-image of target #1, the related data in the frequency/slow-time domain and the target range profiles.

Figure 9.10 shows instead the Radon transform applied to the range profile in Figure 9.9, the IC curve against α_2 and the range profile of the target after motion compensation and the ISAR image in the range/Doppler domain. As can be noted the range profiles have been correctly aligned after the application of the ISAR processing.

Finally, Figure 9.11 shows the results of the cross-range scaling algorithm in terms of chirp rate estimates with the regression LSE straight line, and the ISAR image in the fully spatial domain. More specifically, in Figure 9.11 (a),

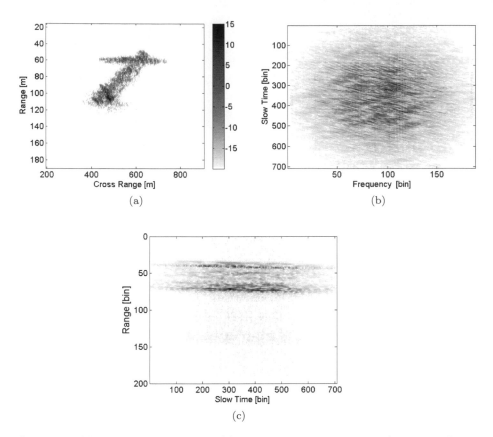

Figure 9.9 SAR sub-image of Ship 1 (a), raw data in the frequency/slow-time domain (b), range profiles

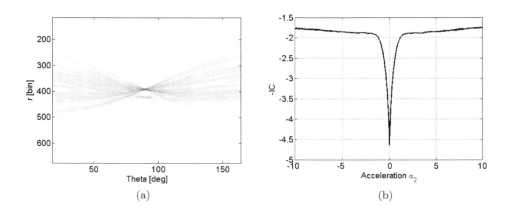

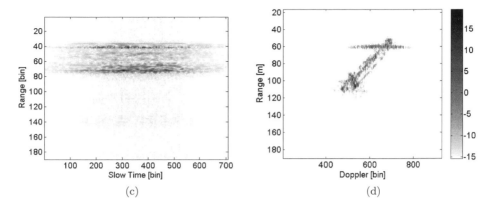

Figure 9.10 Radon transform applied to the range profile in Figure 9.9 (a), IC curve against acceleration α_2 (b),range profiles after motion compensation (c) and ISAR image (d)

the dots represent the chirp-rate estimates of the brightest scattering center composing the targets.

Figure 9.12 shows the results of the ISAR processing applied to target #2, namely, the target sub-image containing cropped from the SAR images, the ISAR images in the range/Doppler domain, the results of the cross-range scaling algorithm in terms of chirp rate estimation and the fully scaled ISAR images.

As can be seen both the results shown the effectiveness of the proposed approach demonstrating that the ISAR algorithm allows for the target motion compensation thus providing an improved image of the target of interest.

The estimated lengths of the ships are $L_{S1} \simeq 75m$ and $L_{S2} \simeq 80m$ for Ship 1 and Ship 2, respectively.

It can be noted, especially in Figure 9.11 (b) an almost straight line aligned with the Doppler/cross-range axis at a given range cell. This appears as a defocusing effect which is likely due to the radar on board of the vessel which generates a micro-Doppler signature because of its rotational motion. Such signature cannot be compensated via the ISAR autofocusing algorithm as ISAR processing relies on the assumption of a rigid-body target.

It should be pointed out that for this dataset the CPI relative to the SAR image is not known. Therefore, the Doppler axis of the ISAR images cannot be given in Hz. However, the effectiveness of the cross-range scaling algorithm does not depend on the CPI and therefore the ISAR images can be represented in the range/cross-range domain. The cross-range scaling algorithm, in other words, estimates the aspect angle variation due to the relative motion between the radar and the target, namely $\Delta\vartheta$. Such aspect angle variation is linked to the modulus of the target effective rotation vector and the observation time

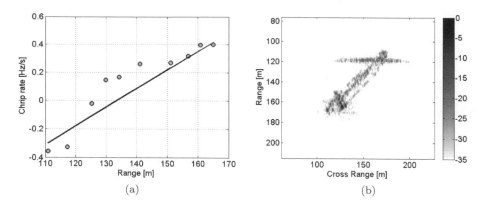

Figure 9.11 Regression straight line (a) and fully scaled ISAR image (b)

via the following relation: $\Delta\vartheta \simeq \Omega_{eff}T_{obs}$. Therefore, whether the CPI is incorrect, the relative estimate of the modulus of the target effective rotation vector will be affected by the same error such that the estimate of the aspect angle variation remains unchanged. Then, even if the information about the target effective rotation vector is incorrect, reliable target sizes are obtained.

As it can be noted the chirp rate estimates fit pretty well the regression straight line. It is worth noting that, although the actual length of the ships is unknown, the estimated ship's sizes obtained by applying the cross-range scaling algorithm are all likely values. Also, targets are projected onto the IPP. Therefore, any estimated distance between two points is the result of a projection, then producing an underestimated value. The ISAR IPP is not a priori known, since it depends on the orientation of the target effective rotation vector, and therefore the projection of the target onto the IPP cannot be a priori determined.

The adopted cross-range scaling algorithm is based on phase information. The results show that the phase information is nicely preserved. This means that the inversion of the unfocused target image back to the spatial frequency domain preserves the phase information.

9.5.2 COSMO SKY-MED: ISTANBUL DATASET

In this sub-section a SAR image acquired by means of COSMO-SkyMed system is considered. Figure 9.13 shows the SAR image covering the area of Istanbul. The circled ships have been processed in this sub-section by means of the ISAR processing.

In this case four targets have been analyzed. For three of them, the time-windowing approach is not necessary.

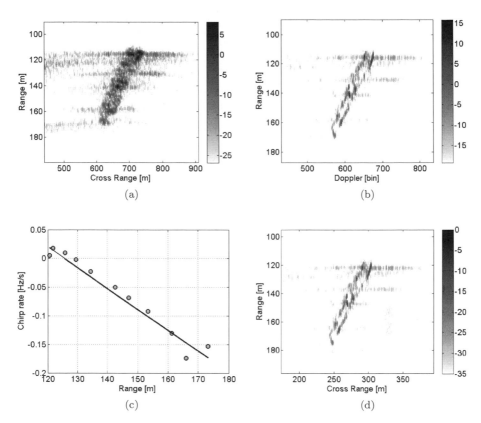

Figure 9.12 SAR sub-image of Ship 1 (a), ISAR image (b), regression line with chirp rate estimates (c), fully scaled ISAR image (d)

Figure 9.14, Figure 9.15, and Figure 9.16 show the sub-images of each target, the corresponding ISAR images in the range-Doppler domain, the results of the cross-range scaling algorithm in terms of chirp rate estimates and the fully scaled ISAR images.

The estimated ships' lengths are $L_{S1} \simeq 66\,m$, $L_{S2} \simeq 51\,m$ and $L_{S3} \simeq 80\,m$ for Ship 1, Ship 2, and Ship 3, respectively.

Also for these cases the chirp-rate estimates fit pretty well the regression straight line, meaning that the estimation of the aspect angle variation and therefore of the cross-range image resolution is trustworthy.

To get well-formed ISAR images of the last target the time-windowing is instead needed. As can be seen from Figure 9.17, the ISAR processing applied to the entire CPI is not able to form a well-focused ISAR image.

Then a time-windowing approach is applied to get a sequence of ISAR images of the target. This algorithm finds the optimal time window length

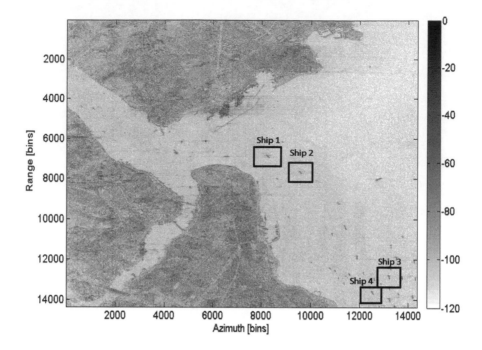

Figure 9.13 SAR image with ships circled

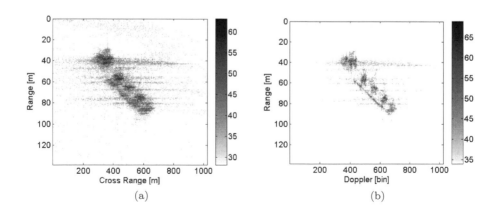

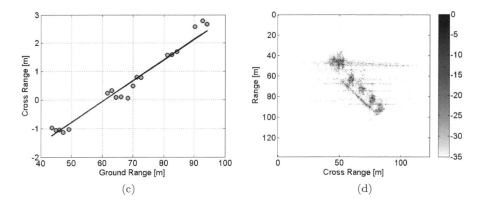

Figure 9.14 SAR sub-image of Ship 1 (a), ISAR image (b), regression line with chirp rate estimates (c), fully scaled ISAR image (d)

for a fixed time instant. The optimality criteria is the following one: take the wider time window that gets a well-formed ISAR image. This criteria uses the image contrast (IC) to measure the quality of the ISAR image against the window time length. Figure 9.18 shows the results of the time-windowing algorithm. It can be noted that these ISAR images differ for the Doppler resolution. This is due to the fact the the integration time (CPI) changes from image to image and is automatically chosen by the same algorithm. As can be noted, the target is manoeuvring as its orientation changes slightly from frame to frame.

9.5.3 COSMO SKY-MED: MESSINA DATASET

Another SAR image acquired by means of COSMO-SkyMed system is analyzed in this sub-section. Figure 9.19 shows the SAR image which covers the area of Messina harbour. The circled ships have been processed by means of the ISAR processing.

Three targets have been analyzed. For none of them the time-windowing approach seems to be necessary. Then, an ISAR image of each target has been obtained by processing the whole SAR image CPI. Figure 9.20, Figure 9.21, and Figure 9.22 show the sub-images of each target, the corresponding ISAR images in the range-Doppler domain, the results of the cross-range scaling algorithm in terms of chirp rate estimates and the fully scaled ISAR images.

The estimated ship's lengths are $L_{S1} \simeq 131\,m$, $L_{S2} \simeq 143\,m$ and $L_{S3} \simeq 80\,m$ for Ship 1, Ship 2, and Ship 3, respectively.

The results in term of chirp rate estimates show that the chirp-rate estimates fit pretty well the regression straight line, meaning that the estimation

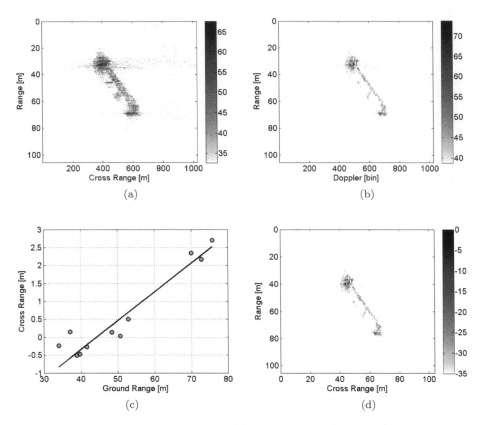

Figure 9.15 SAR sub-image of Ship 1 (a), ISAR image (b), regression line with chirp rate estimates (c), fully scaled ISAR image (d)

of the aspect angle variation and therefore of the cross-range image resolution is trustworthy.

9.5.4   COSMO SKY-MED: SOUTH AFRICA DATASET

Measurements are taken in False Bay, near Cape Town, South Africa. The scenario is a typical coastal maritime surveillance one. The target is a small sailing ship. Some of the spotlight SAR image acquisition parameters are shown in Table 9.1.

The target was slowly sailing, however, significant roll, pitch and yaw motions were induced by the sea waves. Attitude measurements were taken during the acquisition time. A picture of the target is shown in Figure 9.23. The target was a cooperative target and was continuously monitored by a GPS system during the acquisition time.

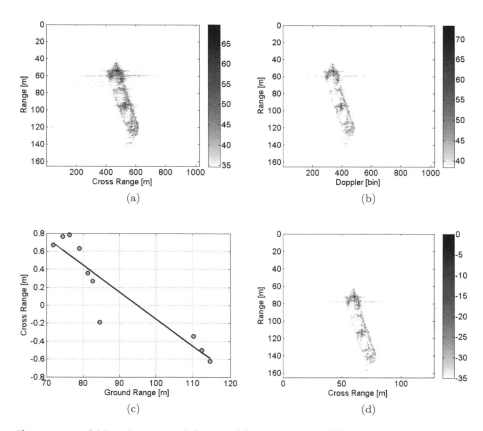

Figure 9.16 SAR sub-image of Ship 1 (a), ISAR image (b), regression line with chirp rate estimates (c), fully scaled ISAR image (d)

The COSMO Sky-Med transmission starting time has been detected by means of a custom acquisition system connected to a spectrum analyzer.

A portion of the SAR image is shown in Figure 9.24 while the sub-image of the target is shown in Figure 9.25. A clear defocusing effect is noticeable due to the sailing ship rapid roll, pitch and yaw motions during the acquisition time. One of such effect is that of showing positive and negative Doppler frequencies caused by the mast oscillation.

As stated in Chapter 3, the hypothesis of quasi constant effective rotation vector within the observation time is one of the hypothesis underlying the ISAR image formation.

In such an experiment, such a hypothesis does not hold because of the target motions which are highly time variant. Then, in order to effectively apply the ISAR algorithm the integration time must be controlled by means of the time windowing algorithm. It must be pointed out that because of the

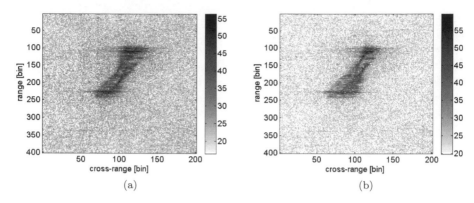

Figure 9.17 SAR sub-image of Ship 4 (a), ISAR image of the same target by processing the entire CPI (b)

Table 9.1
Parameters of interest of the False Bay SAR image

Image size	$10x10\ km^2$
Scene Center	$3413'18.25''\ S\ 1828'38.31''\ E$
Central frequency	$9.6\ GHz$
Range Bandwidth	$220\ MHz$
PRF	$3041\ Hz$
Orbit	Ascending
Look direction	Left (Looking West)
Incidence angle	49 degrees

time dependence of the effective target rotation, the IPPs may change from image to image.

An ISAR image sequence was obtained by windowing the data after the inversion from the image to the data domain. Some of the frames of the ISAR images sequence are shown in Figure 9.26. Such ISAR images sequence has been obtained by selecting a time window of 0.4 s with an overlap of 0.1 s. The effect of the ship roll is evident as the image of the ship mast changes orientation within the observation time.

9.5.5 COSMO SKY-MED DATASET 4: COOPERATIVE TARGET

An experiment with a fully cooperative target has been carried out at the University of Pisa. The objective of such an experiment was to assess the pro-

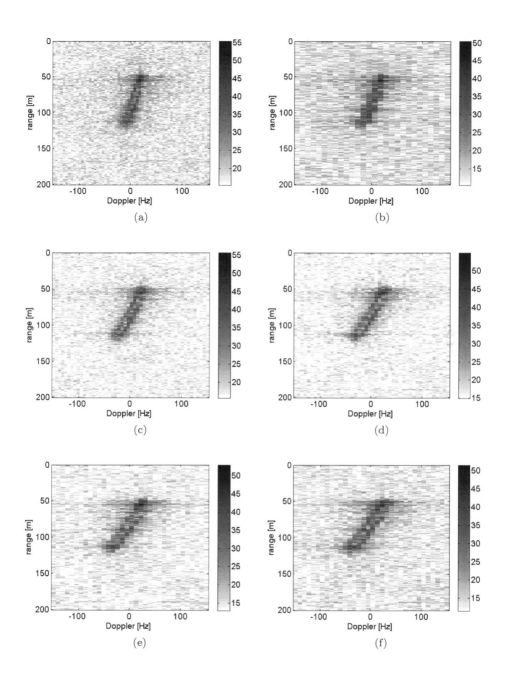

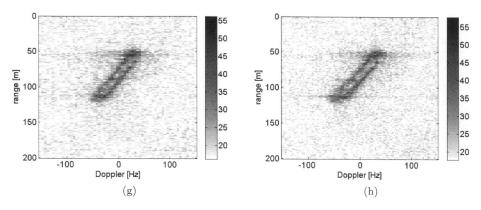

Figure 9.18 ISAR images corresponding to different time instants

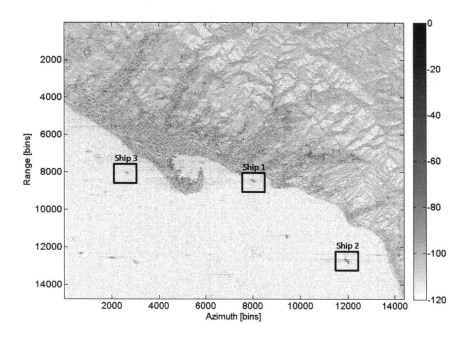

Figure 9.19 SAR image with ships circled

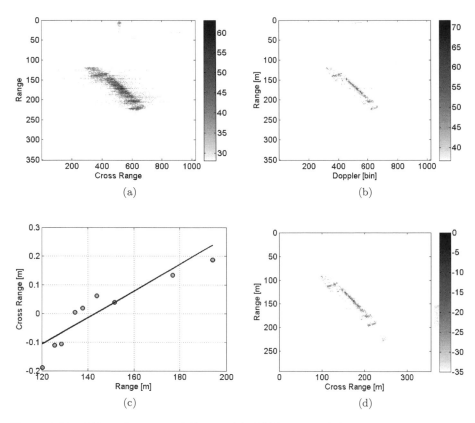

Figure 9.20 SAR sub-image of Ship 1 (a), ISAR image (b), regression line with chirp rate estimates (c), fully scaled ISAR image (d)

cessing chain but more specifically the cross-range scaling algorithm accuracy. A picture of the target is shown in Figure 9.27. The target was a rotating platform composed of four arms at whose extremities there are the quadrihedrals. The quadrihedrals have been realized custom and their radar cross-section has been measured in an anechoic chamber for different incidence angles, specifically $\theta_{in} = [30°, 36° 45°]$. The transmitted signal was a stepped frequency pulsed waveform vertically polarized with carrier frequency $f_0 = 9.6\ GHz$, bandwidth $B = 400\ MHz$ and frequency step $\Delta f = 7.5\ MHz$. After each radar sweep the quadrihedral was rotated by 1°. Figure 9.28 (a) shows the RCS values of the trihedral corresponding to the three incidence angles and the central frequency, namely 9.8 GHz. Figure 9.28 (b) represents instead the RCS values acquired with the incidence angle 30 and for all the frequencies in the frequency bandwidth.

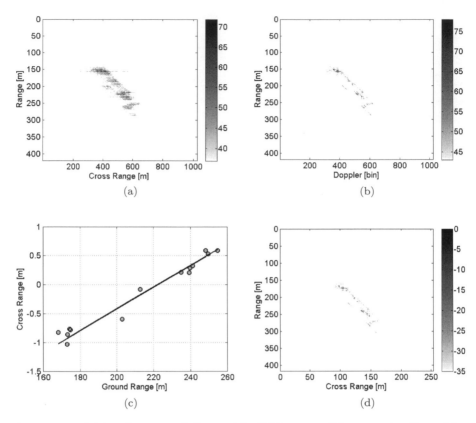

Figure 9.21 SAR sub-image of Ship 1 (a), ISAR image (b), regression line with chirp rate estimates (c), fully scaled ISAR image (d)

The platform maximum extent was 5 m and the platform rotates with a constant platform which can manually set in the interval $[0 \div 0.2]$ rad/s. The angular velocity should be high enough so that the reflectors migrate across several resolution cells within the CPI thus causing both defocusing effects and scatterer displacement in the SAR image. The highest value of the angular velocity is instead dictated by the maximum allowed azimuth displacement. Both the lowest and highest bounds of the angular velocity have been found by means of simulation before the COSMO SkyMed measurements, taking into account the measurement site. The target was located at 43.721099 N 10.383475 E at the Department of Information Engineering, Pisa, Italy. Two measurements have been taken with COSMO SkyMed system, the former on November 9, 2011 and the latter on January 20, 2012.

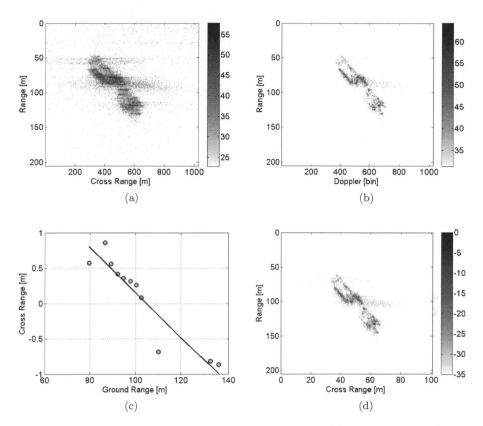

Figure 9.22 SAR sub-image of Ship 1 (a), ISAR image (b), regression line with chirp rate estimates (c), fully scaled ISAR image (d)

9.5.5.1 First Acquisition

Figure 9.29 shows the SAR image covering the area of Pisa (the leaning tower and the church are visible on the top right-hand corner) while Figure 9.30 shows a sub-image of the same SAR image where the target is circled.

For the purpose of comparison, the optical image of the same area is shown in Figure 9.31.

From Figure 9.30 it can be noted that the scatterers undergo an evident shift along the cross-range axis, which depends on both the scatterers' range and cross-range coordinates and the target angular speed. Such an effect, as well as the range and cross-range defocusing effects, prevent us from correctly estimating the target actual size.

Table 9.2 lists some important acquisition parameters.

Once the target has been cropped from the SAR image and projected back to the spatial frequency domain, a sequence of target radar images has been

Figure 9.23   Picture of the cooperative target—sailing ship

Figure 9.24   SAR image of False Bay with the target

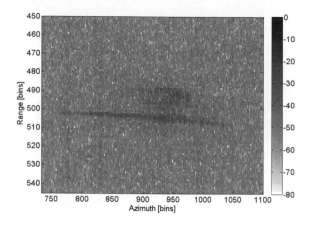

Figure 9.25 SAR sub-image of the target

Table 9.2
Acquisition parameters

f_0	9.6 GHz
B	206 MHz
Synthetic aperture duration	1.92 s
Range spacing	0.425 m
Azimuth spacing	0.701 m
Incidence angle	55°
Pass	Descending
Look side	Right
Target angular speed	0.2 rad/s

Table 9.3
Estimated target size

ISAR image	L_{A1}	L_{A2}
Frame #1	5.28 m	4.66 m
Frame #2	5.33 m	5.22 m
Frame #3	5.42 m	5.28 m
Frame #4	5.68 m	5.22 m
Frame #5	5.3 m	5.16 m
Frame #6	5.73 m	5.17 m
Frame #7	5 m	5.78 m
Frame #8	4.72 m	5.59 m

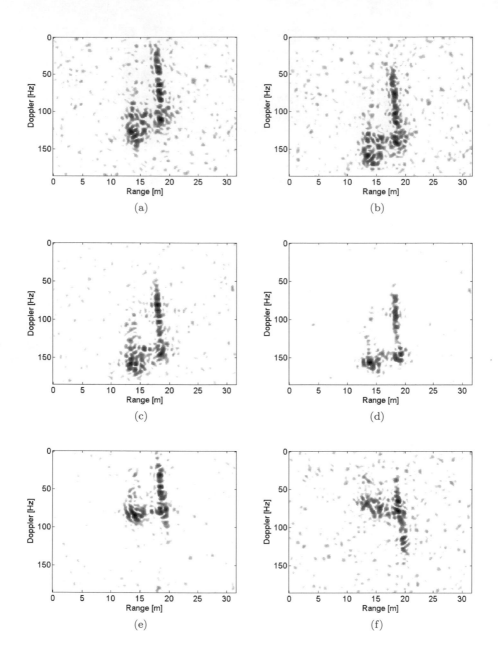

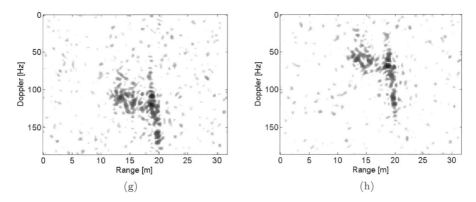

Figure 9.26 A sequence of ISAR images obtained by using a CPI equal to 0.5 seconds. From (a) to (h) it can be noted that the ship's mast moves by changing its Doppler frequencies

Figure 9.27 Target platform where at each end there is a trihedral

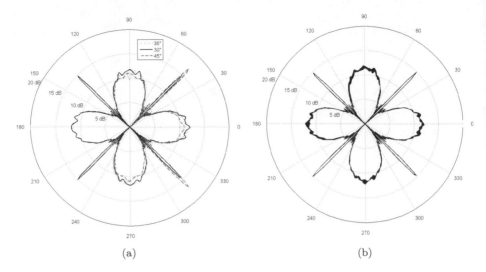

Figure 9.28 RCS values of the quadrihedral measured in an anechoic chamber for different azimuth angles, elevation angles and frequencies. (a) RCS signature for the three elevation angles at the central frequency, (b) RCS signature for different frequencies at a given elevation angle, namely 30°

obtained by using a time moving window of fixed length. Specifically the time window size is 0.3 s and the time interval between two adjacent frame is 0.1 s. Figure 9.32 shows some frames of such sequence, and specifically on the left side there are the ISAR images of the target in a fully spatial coordinate system, while on the right side there are the results of the cross-range scaling algorithm in terms of chirp rate estimate. Such an experiment allows for the performance evaluation of the cross-range scaling algorithm either because the target is cooperative and the IPP is completely known. In fact as the target around x_3 axis, the ISAR IPP conicides with the SAR IPP, namely the slant-range/azimuth plane. Since the orientation of the SAR IPP is completely known, it has been possible to project the ISAR image of the target onto the ground-range/cross-range plane.

As already said, a good indicator of the estimator goodness is the dispersion of the chirp rate estimates around the regression straight line. By observing the results in Figure 9.32, these observations can be drawn:

- The chirp rate estimates are quite concentrated around the regression line.
- The estimated target size from the ISAR image is very close to the actual one.

Figure 9.29 SAR image

Figure 9.30 Crop from the SAR image in Figure 9.29

Figure 9.31 Optical image of the area corresponding to the SAR image in Figure 9.30

The estimated target size, specifically the target arms length, namely L_{A1} and L_{A2} defined as in Figure 9.32, are reported for each ISAR image in Table 9.3.

9.5.5.2 Second Acquisition

Figure 9.29 shows the SAR image covering the area of Pisa (the leaning tower and the church are visible on the top right-hand corner) while Figure 9.34 shows a sub-image of the same SAR image where the target is circled.

From Figure 9.34 it can be noted that the scatterers undergo an evident shift along the cross-range axis, which depends on both the scatterers' range and cross-range coordinates and the target angular speed. Such an effect, as well as the range and cross-range defocusing effects, prevent us from correctly estimating the target actual size.

Table 9.4 lists some important acquisition parameters.

Once the target has been cropped from the SAR image and projected back to the spatial frequency domain, a sequence of target radar images has been obtained by using a time moving window of fixed length. Specifically the time window size is 0.3 s and the time interval between two adjacent frame is 0.1 s. Figure 9.35 shows some frames of such sequence, and specifically on the left side there are the ISAR images of the target in a fully spatial coordinate system, while on the right side there are the results of the cross-range scaling algorithm in terms of chirp rate estimate.

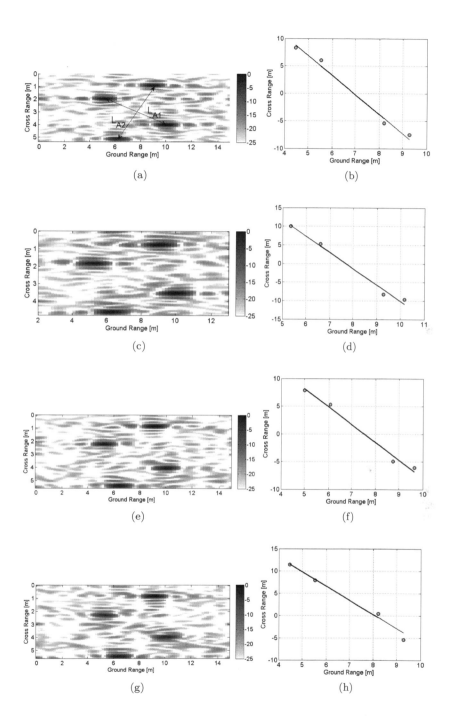

(a)

(b)

(c)

(d)

(e)

(f)

(g)

(h)

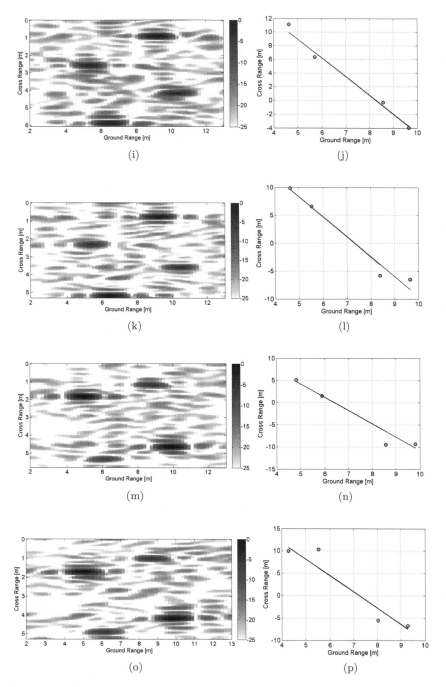

Figure 9.32 Radar images of the rotating platform in the ground-range/cross-range space (left), and results of the cross-range scaling algorithm in terms of chirp rate estimate (right)

Figure 9.33 SAR image

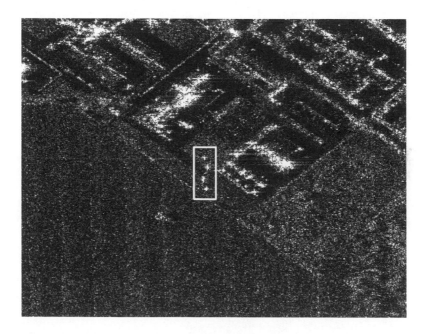

Figure 9.34 Crop from the SAR image in Figure 9.33

Table 9.4

Acquisition parameters

f_0	9.6 GHz
B	225 MHz
Synthetic aperture duration	1.7 s
Range spacing	0.391 m
Azimuth spacing	0.701 m
Incidence angle	48°
Pass	Descending
Look side	Right
Target angular speed	0.1047 rad/s

Table 9.5

Estimated target size

ISAR image	L_{A1}	L_{A2}
Frame #1	5.09 m	4.72 m
Frame #2	4.346 m	3.98 m
Frame #3	4.58 m	4.2 m

The estimated target size, specifically the target arms length, namely L_{A1} and L_{A2} defined as in Figure 9.35, are reported for each ISAR image in Table 9.5.

9.6   ACKNOWLEDGMENTS

The authors thank the Italian Space Agency (ASI) for partially supporting and funding this work under the framework of the first announcement of opportunity for the exploitation of COSMO Skymed products, and for providing COSMO Skymed spotlight SAR images.

REFERENCES

1. Dale A. Ausherman, Adam Kozma, Jack L. Walker, Harrison M. Jones, and Enrico C. Poggio. Developments in radar imaging. *Aerospace and Electronic Systems, IEEE Transactions on*, AES-20(4):363–400, 1984.
2. W.G. Carrara, R.S. Goodman, and R.M. Majewski. *Spotlight Synthetic Aperture Radar: Signal Processing Algorithms*. Artech House signal processing library. Artech House, Incorporated, 1995.
3. C.W. Chen. Modified Polar Format algorithm for processing spaceborne SAR data. In *Radar Conference, 2004. Proceedings of the IEEE*, pages 44–49, April 2004.

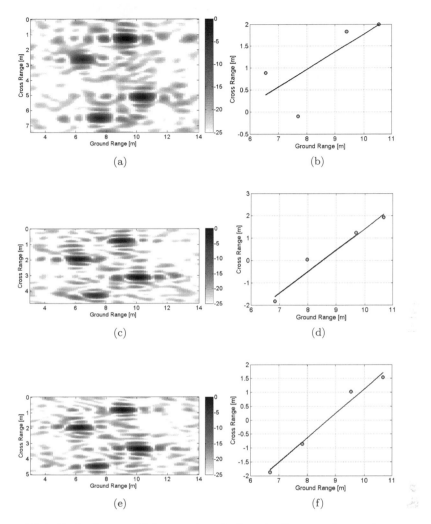

Figure 9.35 Radar images of the rotating platform in the ground-range/cross-range space (left), and results of the cross-range scaling algorithm in terms of chirp rate estimate (right)

4. V.C. Chen and W.J. Miceli. Simulation of isar imaging of moving targets. *Radar, Sonar and Navigation, IEE Proceedings -*, 148(3):160–166, June 2001.

5. J.C. Curlander and R.N. McDonough. *Synthetic Aperture Radar: Systems and Signal Processing*. Wiley Series in Remote Sensing and Image Processing. Wiley, 1991.

6. I. Djurovic, T. Thayaparan, and L.J. Stankovic. SAR imaging of moving targets using polynomial Fourier transform. *Signal Processing, IET*, 2(3):237–246, 2008.

7. A.S. Khwaja, L. Ferro-Famil, and E. Pottier. SAR raw data generation using Inverse SAR image formation algorithms. In *Geoscience and Remote Sensing Symposium, 2006. IGARSS 2006. IEEE International Conference on*, pages 4191–4194, July 2006.

8. M. Kirscht. Detection and imaging of arbitrarily moving targets with single-channel SAR. *Radar, Sonar and Navigation, IEE Proceedings -*, 150(1):7–11, 2003.

9. F. Berizzi A. Bacci E. Dalle Mese M. Martorella, E. Giusti. An ISAR technique for refocusing moving targets in SAR images. In Chen C. H., editor, *Signal and Image Processing for Remote Sensing*. CRC Press, Taylor and Francis Group.

10. M. Martorella. Novel approach for ISAR image cross-range scaling. *Aerospace and Electronic Systems, IEEE Transactions on*, 44(1):281–294, 2008.

11. M. Martorella and F. Berizzi. Time windowing for highly focused ISAR image reconstruction. *Aerospace and Electronic Systems, IEEE Transactions on*, 41(3):992–1007, 2005.

12. M. Martorella, E. Giusti, F. Berizzi, A. Bacci, and E. Dalle Mese. ISAR based techniques for refocusing non-cooperative targets in SAR images. *Radar, Sonar Navigation, IET*, 6(5):332–340, June 2012.

13. A. Moreira and Yonghong Huang. Airborne SAR processing of highly squinted data using a chirp scaling approach with integrated motion compensation. *Geoscience and Remote Sensing, IEEE Transactions on*, 32(5):1029–1040, Sept. 1994.

14. A. Moreira, J. Mittermayer, and R. Scheiber. Extended Chirp Scaling algorithm for air- and spaceborne SAR data processing in stripmap and ScanSAR imaging modes. *Geoscience and Remote Sensing, IEEE Transactions on*, 34(5):1123–1136, Sept. 1996.

15. R.P. Perry, R.C. DiPietro, and R. Fante. SAR imaging of moving targets. *Aerospace and Electronic Systems, IEEE Transactions on*, 35(1):188–200, 1999.

16. R.K. Raney, H. Runge, R. Bamler, I.G. Cumming, and F.H. Wong. Precision SAR processing using chirp scaling. *Geoscience and Remote Sensing, IEEE Transactions on*, 32(4):786–799, July 1994.

17. F.T. Ulaby, R.K. Moore, and A.K. Fung. *Microwave Remote Sensing: Radar remote sensing and surface scattering and emission theory*. Remote Sensing. Addison-Wesley Publishing Company, Advanced Book Program/World Science Division, 1981.

18. Jack L. Walker. Range-doppler imaging of rotating objects. *Aerospace and Electronic Systems, IEEE Transactions on*, 16(1):23–52, Jan. 1980.

19. S. Werness, M.A. Stuff, and J.R. Fienup. Two-dimensional imaging of moving targets in SAR data. In *Signals, Systems and Computers, 1990 Conference Record Twenty-Fourth Asilomar Conference on*, volume 1, pages 16–, 1990.

20. S.A.S. Werness, W.G. Carrara, L.S. Joyce, and D.B. Franczak. Moving target imaging algorithm for SAR data. *Aerospace and Electronic Systems, IEEE Transactions on*, 26(1):57–67, 1990.

21. F. Zhou, R. Wu, M. Xing, and Z. Bao. Approach for single channel SAR ground moving target imaging and motion parameter estimation. *Radar, Sonar Navigation, IET*, 1(1):59–66, 2007.

22. Shengqi Zhu, Guisheng Liao, Yi Qu, Zhengguang Zhou, and Xiangyang Liu. Ground moving targets imaging algorithm for Synthetic Aperture Radar. *Geoscience and Remote Sensing, IEEE Transactions on*, 49(1):462–477, 2011.

10 Passive ISAR for Harbor Protection and Surveillance

*M. Martorella, F. Berizzi, E. Giusti, C.
Moscardini, A. Capria, D. Petri, and M. Conti*

CONTENTS

This chapter presents the application of the PB-ISAR (passive bistatic-ISAR) technique to data collected in a harbor area. The results show the effectiveness of PB-ISAR algorithm for the purposes of harbor protection and surveillance. In Chapter 5, the theory behind PB-ISAR imaging has been detailed. The PB-ISAR algorithm is based on two main concepts. The first concerns the bistatic ISAR theory. Owing to the fact that a passive radar is intrinsically bistatic, since the transmitter and the receiver are not co-located, the bistatic ISAR theory can be applied [4].

The second concept concerns the application of ISAR processing to unfocused range-Doppler images of moving targets. The possibility to use ISAR processing to refocus unfocused SAR images of moving targets has been demonstrated first in [3]. As a SAR image of a moving target may be interpreted as an unfocused range-Doppler image, this becomes perfectly equivalent to what happens when forming range-Doppler maps of moving targets with a PBR system. Because of this analogy between SAR images and RD images, the same processing used in [3] can be used also in the framework of PB-ISAR system.

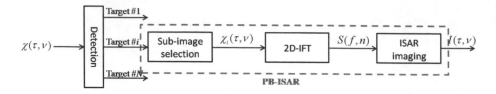

Figure 10.1 Block scheme of the PB-ISAR algorithm

While the theoretical foundation of the PB-ISAR algorithm has been given in Chapter 5, this chapter mainly presents real data results based on the exploitation of DVB-T signals for harbor protection and surveillance.

10.1   PASSIVE BISTATIC ISAR PROCESSING CHAIN

The block diagram of the PB-ISAR algorithm is shown in Figure 10.1.

As shown in Figure 10.1, the algorithm takes as input an RD map, $\chi(\tau, \nu)$. Since it may contain several targets and ISAR techniques can process only a target at a time, the target of interest must be cropped from the RD map so as to isolate it from the other targets, clutter and noise. Since ISAR algorithm usually works with data in frequency/slow-time domain or fast-time/slow-time, the target sub-image should be projected onto one of such data domains. This operation is performed via a two-dimensional inverse Fourier transform, since the RD map is the result of a range-Doppler processing, namely a two-dimensional Fourier transform. Finally, the ISAR processing is applied to such data to get a focused radar image of the target.

In a passive bistatic scenario where both the transmitter and the receiver are stationary, the detection of moving targets is straightforward from an RD map. In fact, differently from stationary targets, the back-scattered signal from a moving target has a Doppler centroid which is proportional to the target radial velocity and is therefore different from 0.

Once a moving target has been detected, the *time windowing enabling* algorithm, proposed in Chapter 9, is applied in this framework.

10.2   PASSIVE ISAR RESULTS

To gather real data, a low-cost passive radar demonstrator has been developed at the RaSS (radar and surveillance system) Laboratory of the University of Pisa, which exploits the DVB-T IOs [1, 5, 2]. The equipment was composed of commercial off-the-shelf low-cost TV antennas and two synchronized Ettus USRP2 boards, which act as signal acquisition boards. The main technical specifications of the USRP2 are listed in Table 10.1. During the experiment the DBSRX was tuned to the central frequency $f_0 = 690$ MHz because three adjacent DVB-T channels centered around f_0 were available. The DBSRX was in fact able to work at lower frequencies (until 600 MHz) with an increase, however, of the noise figure. The equipment was able to acquire a signal bandwidth up to 25 MHz, then, a signal composed of three DVB-T adjacent channels. An example of the baseband spectrum of the acquired signal and of the range ambiguity function are shown in Figure 5.9 and 5.10. The antenna used for the surveillance channel was a 95 element Yagi-Uda with a gain equal to 18 dB. While for the reference channel, a 47 element Yagi-Uda antenna with a gain equal to 15 dB has been used. The measurement was taken near Livorno harbor. Specifically, the transmitter was the DVB-T IO located on Monte Serra, Italy. The receiver was located at the naval academy in Livorno.

Table 10.1
Signal acquisition board specifications

Bandwidth	25MHz
Processing board	FPGA Xilinx Spartan 3-2000 EP1C12 Q240C8 "Cyclone"
Analog digital converter	2 channels
	14 bit
	100 Mega-samples per seconds (100)MS/s
Digital analog converter	2 channels
	16 bit
	400 Mega-samples per seconds (400)MS/s
Interface	Gigabit Ethernet
RF front-ends	DBSRX daughterboard
	Tunable central frequency $f_0 \in [0.8, 2.1]$ GHz

The acquisition geometry is depicted in Figure 10.2, where the receiver, the transmitter and also the BEM are shown. Moreover, the target trajectory, acquired by means of an ADBS receiver, is drawn as well.

Before presenting the results, we should mention again that the television broadcast sources offer quite poor spatial resolutions. This is because IOs usually transmit signals with lower bandwidth and at larger wavelength than those used in dedicated active radar imaging systems. This results in lower range and cross-range resolutions. Moreover, the bistatic geometry degrades both the range and cross-range resolution by a factor $K(0) = K_0 = \cos\left(\frac{\beta(0)}{2}\right)$, and depending on the bistatic angle variation rate during the CPI, it can affect more severely the radar image with distortion effects. To overcome at least

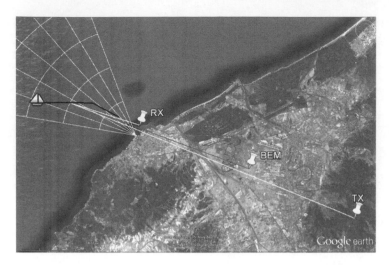

Figure 10.2 Pictorial representation of the acquisition geometry. Image taken from Google Earth.

the limitation on the spatial resolution imposed by the IOs, signals composed of multiple channels and long CPI are required.

Figure 10.3 shows the RD map obtained by means of the cross ambiguity function (CAF) batches algorithm [6]. Four targets are clearly visible in the range-Doppler map. Target 1, which is the closest one to the receiver, was stationary, as it appears at zero-Doppler. Target 2 was approaching the harbor while target Target 3 was leaving the harbor. Target 4 moves slower than the other targets, with respect to the receiver, and it appears at low Doppler frequencies. Because of this, it can hardly be separated from the zero-Doppler clutter ridge. A filter to remove zero Doppler clutter ridge should be applied to remove or at least mitigate the clutter return. For this reason, only Target 2 and Target 3 are considered in the following sections.

The delay time is related to the target-transmitter and target-receiver distances by means of the following equation:

$$\tau = \frac{R_{TxTg} + R_{RxTg} - R_{TxRx}}{c} \tag{10.1}$$

10.2.1   FIRST CASE STUDY

Figure 10.4 shows the range-Doppler sub-image of Target 2. The delay-time axis has been converted to the range axis by using Equation (10.1).

By applying a two-dimensional inverse Fourier transform to the target sub-image, the target echo in the frequency/slow-time domain can be obtained, as shown in Figure 10.5.

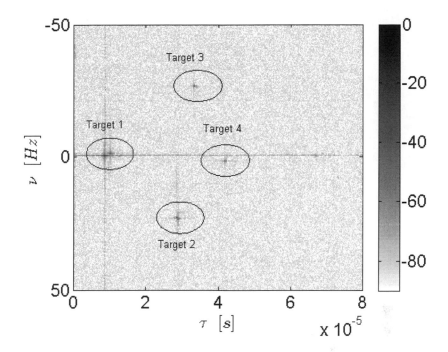

Figure 10.3 Range-Doppler map

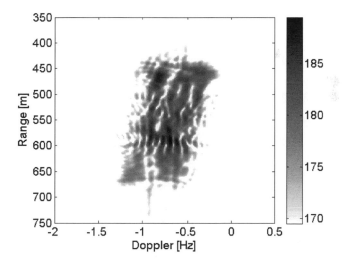

Figure 10.4 Range-Doppler sub-image of Target 2

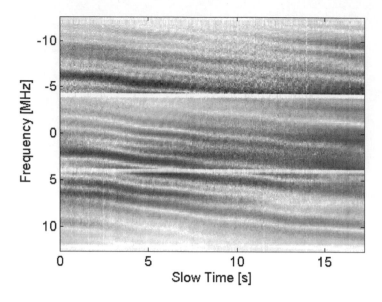

Figure 10.5 Frequency/slow-time data of Target 2

The three adjacent DVB-T channels and the gaps between them are clearly visible. By applying a one-dimensional Fourier transform to the frequency domain, the range profiles of the target can be obtained as shown in Figure 10.6.

Figure 10.6 highlights the need for target motion compensation, as the target migrates across different range cells. The autofocusing algorithm adopted here is the ICBA algorithm. Figure 10.7 shows the Radon transform applied to the range profiles of Figure 10.6, and the IC curve against target radial acceleration.

The range profiles after motion compensation are instead shown in Figure 10.8.

Finally, the ISAR image in the range/Doppler domain is shown in Figure 10.9.

To get an ISAR image in a fully spatial coordinate system, the cross-range scaling algorithm is then applied to the ISAR image in Figure 10.9. The results are shown in Figure 10.10, where both the estimated regression straight line and the ISAR image in the range/cross-range domain are shown. The estimated ship length is about 150 m. The target was a cargo ship of length 212 m. The target length is slightly underestimated. However, it should be pointed out that the estimated target length is affected by a measurement error. Also, the target is projected onto the IPP. Therefore, any estimated distance between two points in the ISAR image is the result of a projection, then producing an underestimated value. The ISAR IPP is not a priori known

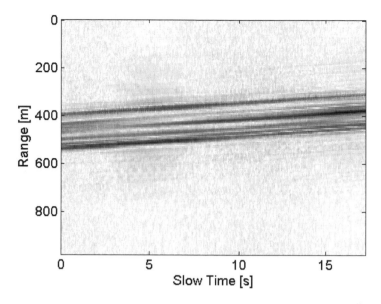

Figure 10.6 Range profiles of Target 2

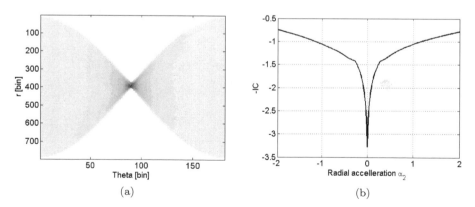

Figure 10.7 Radon transform applied to the range profile in Figure 10.6 (a), and IC curve against target radial acceleration α_2 (b)

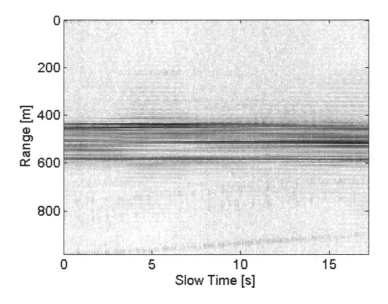

Figure 10.8 Range profiles of Target 2 after motion compensation

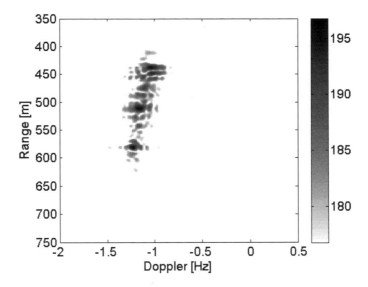

Figure 10.9 ISAR image of Target 2

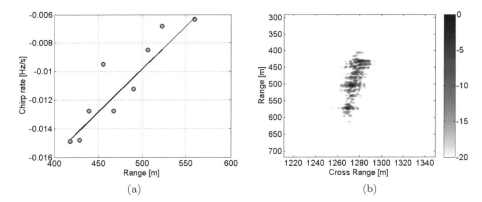

Figure 10.10 Results of the cross-range scaling algorithm in terms of chirp rate estimates (a) and ISAR image in range/cross-range domain

since it depends on target motions, and therefore the projection of the target onto that plane cannot be be a priori determined.

As can be noted, a well-focused image of the target can be obtained by applying the proposed method and this is confirmed by the fact the the radial motion of the target is effectively compensated as shown in Figure 10.8.

10.2.2 SECOND CASE STUDY

Figure 10.11 shows the range-Doppler sub-image of Target 3, and Figure 10.12 shows the ISAR data in the frequency/slow-time domain.

As done previously, the ICBA algorithm has been applied to the whole CPI, giving the results in Figure 10.13, where both the range profiles before motion compensation, the Radon transform, the IC against the target radial acceleration, and the range profiles after motion compensation are shown. It can be noted the range migration experienced by the target is quite well compensated. However, as can be seen by observing Figure 10.14, the ISAR image of the target seems to be not well formed. This is probably because the target underwent complex motions induced by target maneuvers or the state of the sea. Complex motions experienced by the target cause a change of the target effective rotation vector within the CPI, which in turn causes a variation of the IPP within the CPI. The hypothesis of constant target effective rotation vector is instead a requirement for the application of the image formation step and the violation of such hypotheses causes the defocusing effects that are quite visible in Figure 10.14. The time-windowing algorithm is then applied with the aim to reach this requirement and form well-formed ISAR images. A sequence of ISAR images of the target taken at different time instants of the whole observation time is shown in Figure 10.15. By comparing

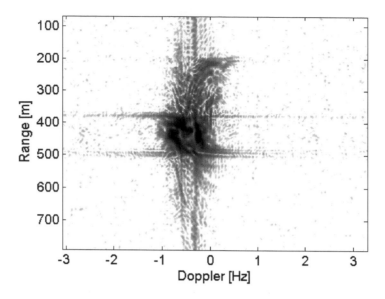

Figure 10.11   Range-Doppler sub-image of Target 3

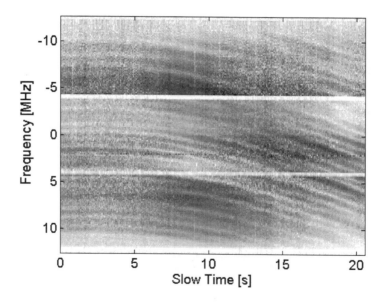

Figure 10.12   Data in the frequency/slow-time domain relative to Target 3

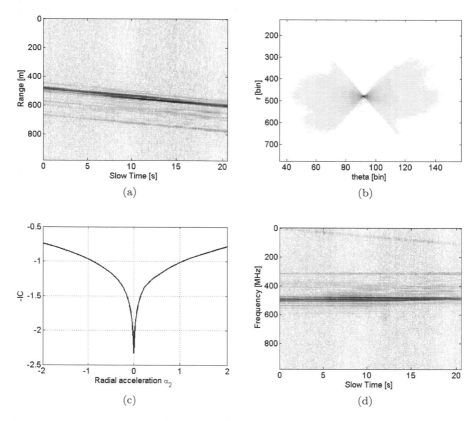

(a)

(b)

(c)

(d)

Figure 10.13 Range profiles before motion compensation (a), Radon transform (b), IC curve against target radial acceleration (c), and range profiles after motion compensation

the ISAR images corresponding to different time intervals, a change in the target's orientation can be observed, which may be due to either the presence of the ship's oscillating motions or to the target's maneuvers, so confirming the hypothesis of variant target effective rotation vector.

10.3 CONCLUSIONS

The theoretical formulation of the PB-ISAR algorithm provided in Chapter 5 has been validated by means of experimental results, thus providing evidence of its capabilities. However, as already mentioned, the broadcast sources offer quite poor spatial resolutions. IOs usually transmit signals with lower bandwidth and larger wavelength than those used in dedicated active radar imaging systems. This results in low range and cross-range resolutions. Moreover, the bistatic geometry degrades both the range and cross-range resolutions by a factor K_0. To overcome this limitation, multiple TV channels and long CPI are required. The first condition is more difficult to achieve because the VHF/UHF is not completely filled by broadcast channels. Then, even if hardware able to acquire a signal with a bandwidth larger than 25 MHz is available, it may be possible that the bandwidth of the acquire signal is not wide as expected or not completely filled. Gaps in the signal spectrum where no signals are transmitted bring to grating lobes and artifacts in the image domain when Fourier transform based algorithms are used to form the ISAR image. Such artifacts dramatically degrade the image quality. To overcome this issue, compressive sensing (CS) theory can be used. CS has been formulated to reconstruct sparse signals by using a limited amount of measures. Results in [7] shows that the proposed CS-ISAR algorithm can be effectively used in the case of spectral gaps.

Finally, it should be remarked that, although passive ISAR images cannot be comparable to active ISAR images in terms of spatial resolutions, passive ISAR imaging remains of great interest because: (i) it provides information on the target RCS at low frequencies where typically active radars cannot transmit, and (ii) as passive radar is intrinsically bistatic, passive ISAR may be a powerful tool to stealthy targets detection and imaging.

10.4 ACKNOWLEDGMENTS

The authors thank the European Defence Agency (EDA) for partially funding this work under the framework of Array Passive ISAR Adaptive Processing (APIS) project.

REFERENCES

1. F. Berizzi, M. Martorella, D. Petri, M. Conti, and A. Capria. USRP technology for multiband passive radar. In *Radar Conference, 2010 IEEE*, pages 225 –229, May 2010.

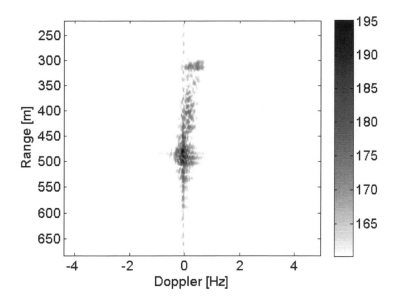

Figure 10.14   ISAR image of Target 3 obtained by processing the entire CPI

2. M. Conti, F. Berizzi, D. Petri, A. Capria, and M. Martorella. High range resolution DVB-T passive radar. In *Proceedings of the European Radar Conference
   (EURAD)*, 2010.
3. M. Martorella, E. Giusti, F. Berizzi, A. Bacci, and E. Dalle Mese. ISAR based
   technique for refocussing non-cooperative targets in SAR images. *IET Radar,
   Sonar, Navigation*, 6(5):1–9, May 2012.
4. M. Martorella, J. Palmer, J. Homer, B. Littleton, and I. D. Longstaff. On
   bistatic inverse synthetic aperture radar. *IEEE Tr on Aerospace and Electronic
   Systems*, 43(3):1125–1134, 2007.
5. D. Petri, F. Berizzi, M. Martorella, E. Dalle Mese, and A. Capria. A software
   defined UMTS passive radar demonstrator. In *Radar Symposium (IRS), 2010
   11th International*, pages 1 –4, June 2010.
6. D. Petri, C. Moscardini, M. Martorella, M. Conti, A. Capria, and F. Berizzi.
   Performance analysis of the batches algorithm for range-doppler map formation
   in passive bistatic radar. In *Radar Systems (Radar 2012), IET International
   Conference on*, pages 1–4, Oct. 2012.
7. W. Qiu, E. Giusti, A. Bacci, M. Martorella, F. Berizzi, H.Z. Zhao, and Q. Fu.
   Compressive sensing for passive isar with dvb-t signal. In *Radar Symposium
   (IRS), 2013 14th International*, volume 1, pages 113–118, June 2013.

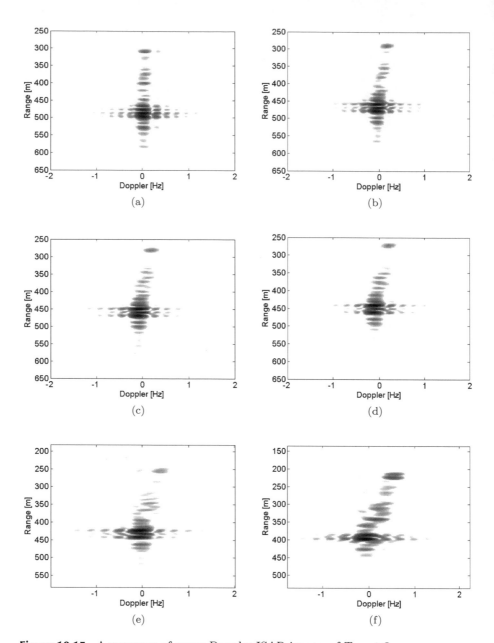

Figure 10.15   A sequence of range-Doppler ISAR images of Target 3

11 Blue Port Traffic Monitoring via 3D InISAR Radar Imaging System

F. Berizzi, M. Martorella, E. Giusti, F. Salvetti,
D. Staglianò, and S. Lischi

CONTENTS

Nowadays, maritime traffic control is mainly performed via the VTS (vessel traffic service) system, which exploits information from AIS (automatic identification system) systems installed on board of cooperative vessels, remote radar site (RRS) and GPS system. The AIS allows the port authorities to know with adequate accuracy absolute geographic coordinates, velocity, trajectory, type, size, draught and cargo of each cooperative vessel (equipped with AIS) thus providing the means for controlling and monitoring the area of interest.

Maritime traffic control systems are usually equipped with a ground segment, which consists of a coastal surveillance radar. This is able to detect all vessels, both cooperative and non-cooperative, and to provide an "all weather" and "all day" monitoring of a large area of interest.

A modern VTS integrates all this information for effective traffic organization and communication with the law enforcement authority.

Nevertheless, a VTS system may not always be effective.

In fact, AIS systems provide detailed information only of cooperative targets. Conversely, coastal surveillance radars provide detection and tracking of both cooperative and non-cooperative targets. However, it may suffer from blind velocity, its performance may be affected by target scintillation and they cannot provide information about target size, cargo, draught and so on.

By taking into account these drawbacks and the continuous increase of the maritime traffic, modern maritime traffic control systems are demanded to strengthen the maritime traffic control so as to increase the level of security in coastal areas without greatly affecting the e.m. pollution level. The use of several low power active radars as well as spatially distributed passive radars could represent a suitable response to such a demand.

The joint use of the information from each radar of the system permits:

- Reducing the shadowed areas and thus improving the detection capability of the system
- Providing an enhanced robustness against bot target scintillation and blind velocities
- Improving the detection and tracking performance by jointly exploiting the detections from all the radars.

Moreover, if such radars are also designed to have imaging capabilities, the joint use of either the received data and the radar images could provide further and important information. This information can be used to assist the docking maneuvers, to improve the monitoring of illegal activities, to enable ATC/ATR capabilities and for "collision avoidance."

This motivates the need for a deeper investigation of the use of a radar sensor with imaging capability (2D/3D) for harbor surveillance.

The chapter is organized as follows. Section 11.1 presents a high level description of how a distributed multi-sensor network could be conceived. Section 11.2 presents the signal processing chain relative to a radar network operating in a MIMO configuration. Section 11.3 presents experimental results obtained by using a single InISAR radar system and a cooperative target. Finally, conclusions are drawn in Section 11.4.

For a good comprehension of this chapter, the reader should be familiar with 2D and 3D ISAR imaging. Then, a careful reading of Chapters 3, 4, and 6 is recommended.

11.1   SYSTEM ARCHITECTURE

The maritime traffic control system could be conceived as composed of three sub-systems, as shown in Figure 11.1:

1. The harbor navigation system
2. The data processing and communication network
3. The sensor network

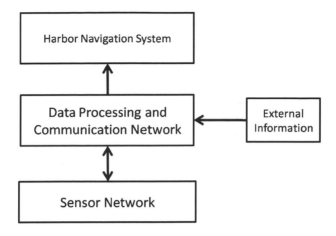

Figure 11.1   Maritime traffic control system

The *sensor network* includes radar sensors with imaging capability since they are "all vessels, all weather, all day" sensors, designed to operate in a MIMO (multiple input multiple output) configuration. To facilitate their transportation and installation in a complex environment as a harbor, low-power, low-weight and compact radars are recommended. Radar sensors satisfying such requirements are both passive radar systems exploiting illuminator of opportunities (IOs), such as television signals, and active radars based on FMCW technology. In fact, differently from pulse radars which require high peak transmission power, FMCW radars operate with constant low power, which means lower cost and smaller size [2, 1]. Obviously, other systems could cooperate with the radar sensors, such as electro-optical (EO), infrared (IR) and LIDAR sensors to further improve the system accuracy and capability. However, in this chapter, the focus is given to imaging radar sensors.

The *data processing and communication network* is meant to manage the communication among radar sensors, the transmission of data to and from a ground control station, and from the ground control station to devices on board of vessels. It is also meant to manage the information coming from each radar sensor and to implement algorithms such as multistatic detection, tracking and imaging, which are devoted to enable navigation services like collision avoidance and docking facilitation applications. Finally, the *harbor navigation system* implements navigation services such as collision avoidance and docking maneuvers. A pictorial representation of the maritime control system architecture is shown in Figure 11.1.

11.2   SYSTEM'S SIGNAL PROCESSING CHAIN

Let N_S be the number of imaging ground-based radars composing the sensor network. A block diagram of a possible multistatic signal processing algorithm is depicted in Figure 11.2.

Here, it has been supposed that N_{tg} targets are located in the area of interest.

As a first step of the processing chain, each radar performs a local detection onto the range-Doppler (RD) map. Each radar detects a certain number of targets, namely, N_{tg_s}, where $s = 1, \cdots, N_S$. Because of shadowing effects and target scintillation, it may happens that $N_{tg_s} \leq N_{tg}$.

However, as each sensor detects targets in the range-Doppler domain, targets cannot be localized in a 3D spatial domain (geographical coordinates). Then, a multistatic detector is applied which processes the detection results of each radar sensor and performs multilateration and target association to give the Cartesian coordinates of each target in the scene. At least three radars should detect the target to geolocate it without any ambiguity. Before performing multilateration, however, data association must be performed. Data associations deals with the problem of selecting the measurements (range-Doppler cells occupied by a target, target velocity, etc.) originated from the same targets.

The multistatic detector gives as output each target position vector, $\mathbf{x}_g$, and velocity vector, $\mathbf{v}_g$, where $g = 1, \cdots, N_{tg}$. The parameter N_{tg}, as already said, is the number of targets detected by the radar sensor network.

After the multistatic detection step, each target is processed by each radar imaging sensor. The radar imaging sensor could be designed so as to be able to produce either a 2D ISAR image or a 3D reconstruction of the detected target. The *2D/3D radar imaging* block processes the RD maps from each radar which has detected the target of interest, thus producing n_S 2D or 3D images of the targets. n_S represents the number of radars which have detected the target under test and is usually lower than the number of radar sensors, that is $n_S \leq N_S$. 2D ISAR images and 3D reconstructed models of the target can be obtained from RD maps by means of algorithms reported in Chapter 3 and Chapter 6.

Finally, the *multistatic radar imaging* block aims at fusing together the results of the radar imaging processing at each sensor with the aim to provide an improved 2D ISAR image or a more complete 3D reconstructed model of the target as well as to give further information about the target such as its size and shape. Such information together with its velocity and direction could be used for both *collision avoidance* and *docking facilitator* applications.

A net of distributed radars has a number of advantages over a monostatic radar: (i) robustness over target aspect angle changes, (ii) robustness over shadowing effects, (iii) target rotation vector estimation capability and (iv) an improved detection and tracking capability. A pictorial representation of a

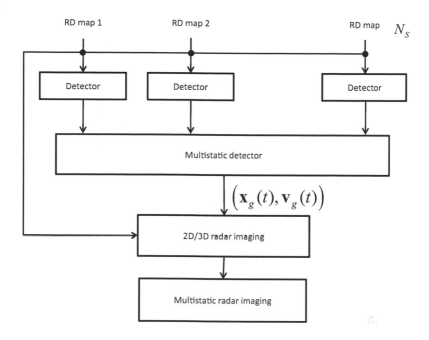

Figure 11.2 Signal processing chain

multistatic configuration composed of widely spatial distributed radar sensors is depicted in Figure 11.3.

However, the big issue that should be addressed is how to combine the data acquired from each radar. Two possible solutions to this problem could be considered which differ in the adopted fusion algorithm. Specifically, data fusion can be performed either at the raw-data level or at the image level. When the fusion is performed at the raw-data level, the received signals are coherently combined in the spatial frequency domain, namely the frequency/slow-time domain. Although this represents the optimal approach, a number of critical aspects have to be considered: (i) the time and phase synchronization among radars should be kept highly accurate, (ii) both the target aspect angle and the target rotation vector should be a priori known or estimated to map the signal in the 3D Fourier space, namely the 3D spatial frequency domain, (iii) since the acquired data depends on the relative position of the maneuvering target with respect to the radar, the data in the Fourier domain could be incomplete (namely, the spatial frequency region where no signal is received), thus affecting the 3D ISAR image when a Fourier-based algorithm is used.

Conversely, when the fusion is performed at the image level, the outputs of each radar processing are incoherently combined in the image domain. This method is less accurate with respect to the others. However, it is less sensitive to phase coherence and errors in the estimation of the target motion, and

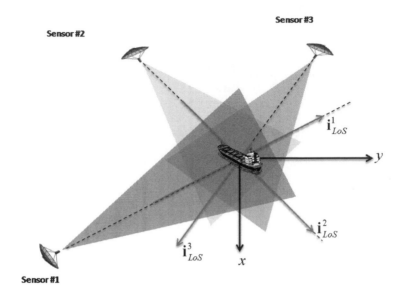

Figure 11.3   Multistatic configuration

therefore it results more feasible especially when radars are very far apart. Nonetheless, in the incoherent image fusion, the fusion of 2D ISAR images is not straightforward. With reference to Figure 11.4, the radars produce two ISAR images. Each ISAR image is the projection of the 3D target reflectivity function onto different 2D planes, arbitrarily oriented in the 3D space. From Figure 11.4, it is possible to understand that such ISAR images cannot be directly combined. In fact, the IPPs need to be oriented in the same way to carry out a 2D fusion. To overcome such a limitation, an interferometric ISAR system can be used instead of a conventional ISAR system. In InISAR systems, the interferometric phase measured by two orthogonal baselines are used to jointly estimate the effective target rotation vector (both in modulus and phase) and the heights of the scattering centers. Then, when such sensors are used, the incoherent summation requires the alignment of the reconstructed 3D target models. Some examples of the reconstructed 3D target models alignment procedure can be found in [3].

11.3   EXPERIMENTAL RESULTS

A measurement campaign has been carried out at the Navy facilities in Livorno, Italy, with the aim to prove the 3D InISAR algorithm presented in Chapter 6 and to demonstrate the usefulness of imaging radar (in this case, InISAR system) for navigation services. Measurements have been carried out using the PIRAD radar, a ground-based radar prototype designed and built by

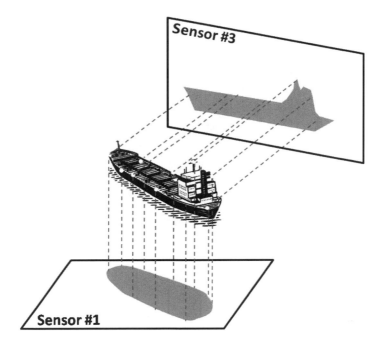

Figure 11.4   ISAR IPPs

Italian radar researchers at both the Department of Information Engineering at of the University of Pisa and at the RaSS (radar and surveillance system) National Laboratory in Pisa. Both cooperative and non-cooperative targets were present during experiments. However, focus will be given in the following subsections to the cooperative target since its size, geometrical characteristics, position and velocity were known. The cooperative target was equipped with a differential GPS system so as to measure its velocity and position over time. Results in terms of 2D ISAR images as well as 3D reconstructed model of the cooperative target are shown subsequently.

11.3.1   EXPERIMENTAL SET-UP DESCRIPTION

PIRAD is a ground-based radar prototype designed and built by Italian radar researchers at both the Department of Information Engineering of the University of Pisa and at the RaSS National Laboratory.

A picture of the PIRAD radar is shown in Figure 11.5. PIRAD has been designed for all-day and all-weather monitoring of ship traffic. PIRAD is a low power and small size sensor based on a linear frequency modulated continuous wave (LFMCW) technology. LFMCW is the most suitable technology for such an application. This type of waveform is able to give information about both distance and radial target velocity. This is achieved by using a quite low

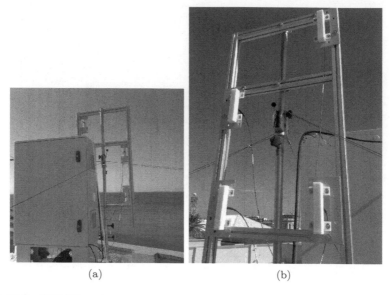

(a)   (b)

Figure 11.5   PIRAD system

radiated power level to cover a range of a few kilometers. Since PIRAD is a CW radar and hence a coherent radar, it can be used for high-resolution imaging purposes.

The main system requirements are summarized in Table 11.1.

The antenna system consists of one transmitting and three receiving elements arranged in an L-shaped frame exactly as shown in Figure 11.5. An X-band Fabry-Perot resonator technology was adopted in order to realize compact and efficient horizontally polarized antennas. The antenna is shown in Figure 11.6. The developed antennas are characterized by the total absence of side lobes on the azimuth plane. The LFMCW transmitted signal is generated

Table 11.1

PIRAD system requirements

Range coverage	1200 m
Electric field intensity	< 6 V/m
SNR after DSP	≥ 20 dB
Range resolution	0.5 m
Elevation angle	$20°$
Azimuth coverage	$60°$
Receiving channels	3
Target radar cross-section	$1 \div 10^4$ m^2

Figure 11.6 PIRAD antenna

Table 11.2
PIRAD transceiver specifications

TX power	up to 33 dBm
Chirp rate	up to 1.5 THz/s
Noise figure	4.2 dB
System losses	12.5 dB
Operative frequency	$10.55 \div 10.85$ GHz
Dynamic range	$-132 \div -38$ dBm
SFDR	65 dBc
Receiver gain	29 dB
ADC resolution	14 bit

by using a phase locked loop (PLL) as a frequency synthesizer. The PLL is digitally driven by a micro controller connected to a personal computer. The receiver architecture is a low-IF direct deramping with radio-frequency analogical beat. The signal acquisition is performed through two linked USRP2. These two USRP2 are connected to a general purpose PC through a gigabit Ethernet link.

Finally, the specifications of the PIRAD transceiver are summarized in Table 11.2.

Experiments have been carried out in Livorno (Italy) at the Italian Navy facilities. The Naval Academy where the radar was set up is near the Livorno harbor. Besides the cooperative target, other targets were present in the scene, namely, cargo vessels almost stationary or slowly approaching the harbor and small sailing ships. However, cargo vessels were at a range out of the radar range coverage. A picture of the scenario taken from Google Earth is depicted

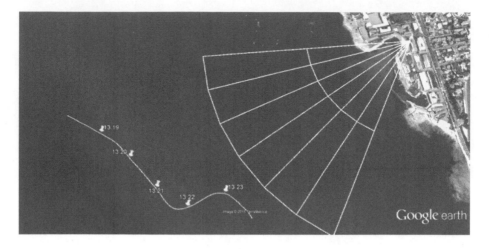

Figure 11.7 Acquisition geometry and GPS data of the cooperative target

Table 11.3
Target sizes

Length	33.25 m
Width	6.47 m
Deckhouse	$\simeq$ 4.5 m
Main mast height	$\simeq$ 8 m

in Figure 11.7 were both the radar position, the antenna azimuth coverage and the cooperative target GPS position over time are represented.

The cooperative target is shown in Figure 11.8 and its sizes are reported in Table 11.3.

From Figure 11.8, it can be noted that the target was equipped with the main mast where radar, antennas and other sensors were in place. This contributes to make the actual height of the target of about 12.5 m.

Finally, a photo of the scene taken almost simultaneously with measurements is represented in Figure 11.9, showing the presence of other targets.

11.3.2 RESULTS

In this section, results of both the conventional 2D ISAR processing applied to the RD map and and results of the 3D InISAR technique are reported to show the effectiveness of both the techniques in providing support for maritime traffic monitoring and harbor surveillance.

The backscattered signal from both the sea clutter and the targets was acquired for $T_{obs} \simeq 130$ s. However, for the sake of simplicity, the results

Figure 11.8 Cooperative target

Figure 11.9 Photo of the scene taken almost simultaneously with measurements

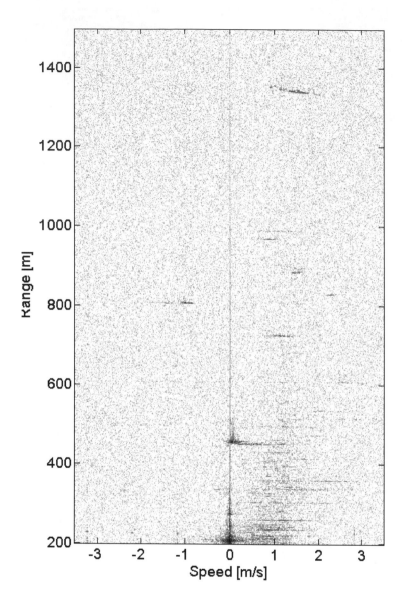

Figure 11.10 RD map relative to the central antenna of the PIRAD system and an integration time equal to $T_{int} \simeq 0.8$ s

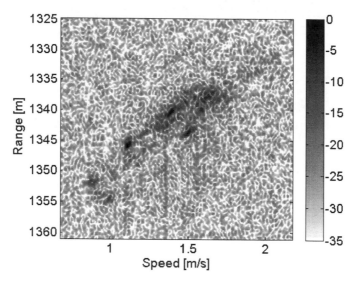

Figure 11.11 Target sub-image extracted from the RD map of the central channel shown in Figure 11.10

obtained by processing only a portion of such data will been shown. Figure 11.10 shows the RD map relative to an integration time equal to $T_{int} \simeq 0.8$ s. Such RD map is relative to a receiving antenna of PIRAD system, namely the central one.

The cooperative target was located at about 1350 m from the radar. Other targets were present in the scene, specifically, a stationary sailing ship at about 450 m and other sailing ships between 700 m and 1000 m far from the radar.

Once the target is detected, a sub-image of the target is extracted from the RD map of each receiving channel. In Figure 11.11 the sub-image of the target extracted from the RD map of the central channel is represented.

After that, the raw-data like relative to the target of interest are obtained by means of a $2D\text{-}IFT$ and then the 2D ISAR processing based on the ICBA algorithm is used. For the sake of simplicity, only the 2D ISAR image relative to the central channel is shown in Figure 11.12, since the ISAR image relative to the other channels is quite similar. In Figure 11.12, the white dots represent the scatterers extracted from the 2D ISAR image by using the multi-channel CLEAN algorithm described in Chapter 6.

Finally, the results of the 3D-InISAR algorithm are shown in Figure 11.13 and Figure 11.14. Specifically, in Figure 11.13 the target model in the 3D space is represented while Figure 11.14 shows the top view and a side view of the image in Figure 11.13. The 3D reconstructed model of the target is represented in the Cartesian coordinate system T_{ξ_r} which is obtained from the Cartesian reference system T_ξ defined in Chapter 6 by means of a rotation, as follows:

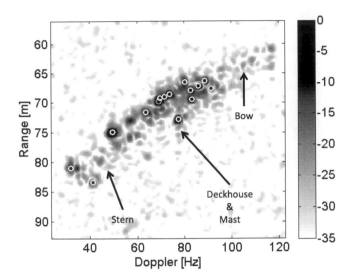

Figure 11.12 2D ISAR image relative to the central channel. White dots represent the scatterers extracted by using the MC-CLEAN technique

$$\xi_r = \mathbf{M}_{rot} \cdot \xi \tag{11.1}$$

where $\mathbf{M}_{rot}$ is a 3D rotation matrix which rotates the target of the azimuth and elevation angles which identify the target velocity vector estimated from the GPS data.

As can be seen by observing Figure 11.12, only some parts of the target can be imaged and then extracted by means of the multi-channel CLEAN algorithm, namely scatterers of the stern, deck-house and mast. The height of such scatterers is then calculated by exploiting the 3D InISAR algorithm, as shown in Figure 11.13 and Figure 11.14.

Other results have been obtained by using the same data but at different time. Specifically, results in Figures 11.15, 11.16, and 11.17 have been obtained by processing a portion of data corresponding to 4 seconds later than that used previously. Conversely, results in Figures 11.18, 11.19, and 11.20 have been obtained by processing a piece of data corresponding to 5 seconds later on. From Figure 11.15 it can be seen that the MC-CLEAN extracts scatterers mainly from the stern, the deck-house and a scatterer from the bow. The MC-CLEAN technique extracts the brightest scatterers, which correspond to the most affordable and coherent ones for the 3D target model reconstruction.

As expected, the scatterers belonging to the deck-house are higher than those ones which belong to the stern or the bow. Moreover, by taking a closer

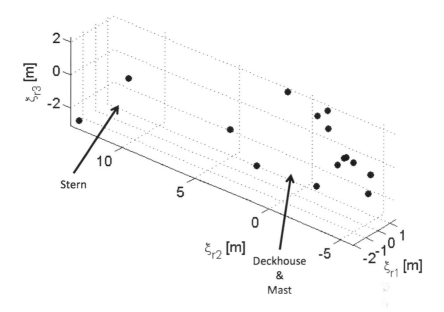

Figure 11.13 3D reconstructed model of the target obtained using the 3D InISAR algorithm and exploiting the extracted scatterers in Figure 11.12

look at the target picture in Figure 11.8, it can be noted that the ship is also equipped with a mast where radar, other sensors and antennas are installed. The mast is quite higher than the deck-house. In the 3D reconstructed target model in Figure 11.16 a scatterer can be identified which has a height ($\xi_{r3} \simeq$ 4.2 m) higher than the others in the desk-house. Then, this scatterer can be associate with a structure of the mast. Furthermore, from the picture in Figure 11.8, it can be seen that the ship is also equipped with two cranes at the stern. In Figure 11.16 and especially in Figure 11.17(b) two scatterers of the stern have been extracted. One of those scatterers has a height of almost 2 m higher than the ship desk, which likely corresponds to a crane.

The 2D ISAR image in Figure 11.18 is the one with the highest energy and then the number of extracted scatterers is larger than in the other cases. Scatterers belonging to the most important parts of the ship, namely, the stern and probably the cranes, the desk-house as well as the bow have been extracted via the MC-CLEAN technique. As can be seen by observing Figure 11.19 and Figure 11.20, the scatterers which are located in the center of the ship that most likely correspond to the deck-house or the mast have a height higher than the others. The scatterers at the stern could represent both a crane or the deck. The scatterer at the bow is a little higher than expected,

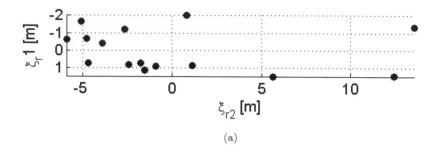

(a)

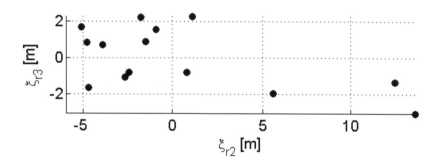

(b)

Figure 11.14   Top view and side view of the 3D target model in Figure 11.13

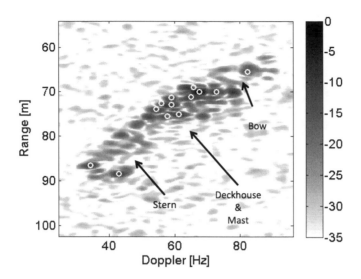

Figure 11.15 2D ISAR image with scatterers extracted via the MC-CLEAN algorithm

but however, if we take a look at Figure 11.8, it can be noted a sort of jibboom or more simply a rod at the bow, which has a height higher than the deck. It is worth noting that the results obtained by processing different time intervals are quite different although the processed time intervals were quite close to each other. Such differences may be due to both target scintillation and target complex motions induced by the sea state.

Then, the exploitation of different time intervals even if from the same radar can provide a better understanding of the target itself. To better demonstrate this concept, the reconstructed 3D target models in Figure 11.13, 11.16 and 11.19 have been aligned by means of affine transformations (translations and rotations). The results of this alignment are reported in Figure 11.21 and Figure 11.22.

As can be seen by comparing Figure 11.21 with Figure 11.13, 11.16 and 11.19, the multi-temporal 3D target model in Figure 11.21 contains much more information about the target thus providing a better understanding of the target. This is because, even if similar, the 3D target models obtained at different time instants add information with respect to the others. Moreover, with such a procedure, persistent scatterers over time could be identified which can add further information about the target. Some of the persistent scatterers have been circled in Figure 11.22(b).

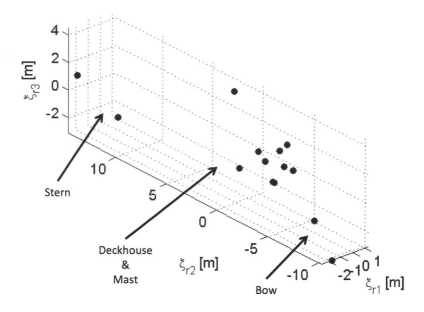

Figure 11.16 3D reconstructed model of the target obtained using the 3D InISAR algorithm and exploiting the extracted scatterers in Figure 11.15

Finally, if we compare the images in Figure 11.21 with the picture of the target in Figure 11.8 a very close relation can be found both in terms of target sizes and shape.

11.4 CONCLUSIONS

The main objective of this chapter was to shown how a radar sensor with 3D imaging capability could be used to monitor and control the maritime traffic inside a harbor.

A measurement campaign has been carried out near the Livorno harbor at the Italian Navy facilities with an InISAR radar. The target was a cooperative target in this case; however, once validated, such an algorithm can be used to image non-cooperative targets. Results have demonstrated the validity of the 3D InISAR that has been proposed in Chapter 6 in estimating the actual target size in the 3D space with high accuracy as well as the target shape. This information can be used for several applications such as to improve the "collision avoidance" system performance, assist the docking maneuvers, improve the monitoring against illegal activities, enable ATC/ATR capabilities

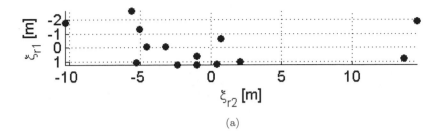

(a)

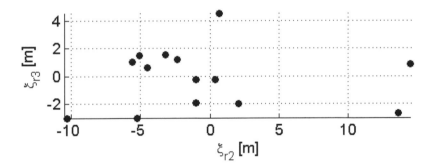

(b)

Figure 11.17 Top view and side view of the 3D target model in Figure 11.16

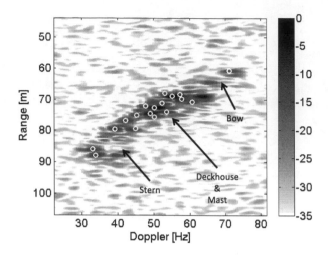

Figure 11.18 2D ISAR image with scatterers extracted via the MC-CLEAN algorithm

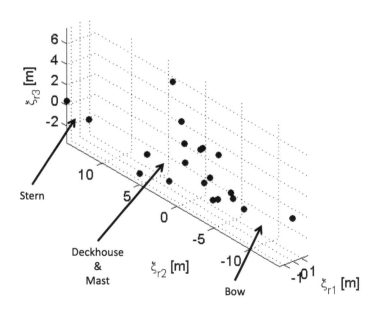

Figure 11.19 3D reconstructed model of the target obtained using the 3D InISAR algorithm and exploiting the extracted scatterers in Figure 11.18

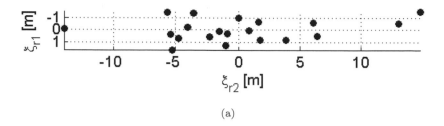

(a)

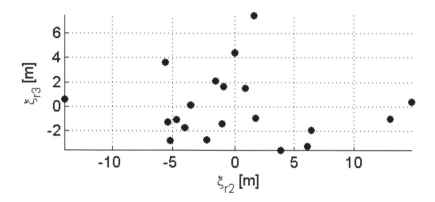

(b)

Figure 11.20 Top view and side view of the 3D target model in Figure 11.19

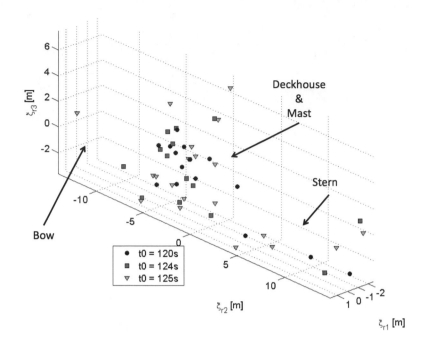

Figure 11.21 3D reconstructed models after alignment

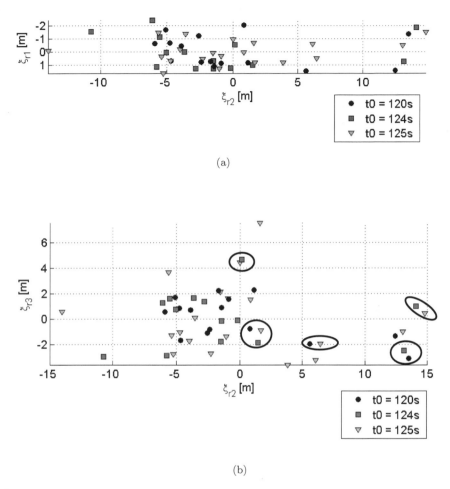

(a)

(b)

Figure 11.22   Top view and side view of the 3D target model in Figure 11.21

and so on. A multi-temporal approach is also described to show that, with a single radar, the use of data corresponding to different time intervals improve the understanding of the target itself. It is quite obvious that the use of several InISAR radar systems properly located throughout the area of interest can further improve the accuracy in the estimation of the target size as well as in target recognition thus opening the door to ATC or ATR applications.

11.5   ACKNOWLEDGMENTS

The authors thank the Italian Ministry of Education University and Research (MIUR) for partially supporting this work in the framework of the *HArBour TraffIc OpTmizAtion sysTem*, HABITAT, project.

REFERENCES

1. Yue Liu, Yun Kai Deng, R. Wang, and O. Loffeld. Bistatic fmcw sar signal model and imaging approach. *Aerospace and Electronic Systems, IEEE Transactions on*, 49(3):2017–2028, July 2013.
2. A. Meta, P. Hoogeboom, and L.P. Ligthart. Signal processing for fmcw sar. *Geoscience and Remote Sensing, IEEE Transactions on*, 45(11):3519–3532, Nov. 2007.
3. F. Salvetti, D. Stagliano, E. Giusti, and M. Martorella. Multistatic 3d isar image reconstruction. In *Radar Conference (RadarCon), 2015 IEEE*, pages 0640–0645, May 2015.

Index

Printed and bound by CPI Group (UK) Ltd, Croydon, CR0 4YY

22/10/2024

01777623-0018